TSA Past Paper Worked Solutions

Volume One

Copyright © 2018 *UniAdmissions*. All rights reserved.

ISBN 978-1-912557-27-1

No part of this publication may be reproduced or transmitted in any form or by any means, electronic or mechanical, including photocopying, recording, or by any information retrieval system without prior written permission of the publisher. This publication may not be used in conjunction with or to support any commercial undertaking without the prior written permission of the publisher.

Published by *RAR Medical Services Limited*
www.uniadmissions.co.uk
info@uniadmissions.co.uk
Tel: 0208 068 0438

TSA is a registered trademark of Cambridge Assessment, which was not involved in the production of, and does not endorse, this book. The authors and publisher are not affiliated with TSA. The answers and explanations given in this book are purely the opinions of the authors rather than an official set of answers.

The information offered in this book is purely advisory and any advice given should be taken within this context. As such, the publishers and authors accept no liability whatsoever for the outcome of any applicant's TSA performance, the outcome of any university applications or for any other loss. Although every precaution has been taken in the preparation of this book, the publisher and author assume no responsibility for errors or omissions of any kind. Neither is any liability assumed for damages resulting from the use of information contained herein. This does not affect your statutory rights.

TSA Past Paper Worked Solutions

Volume One

Joseph Nelson

Rohan Agarwal

About the Author

Rohan is the **Director of Operations** at *UniAdmissions* and is responsible for its technical and commercial arms. He graduated from Gonville and Caius College, Cambridge and is a fully qualified doctor. Over the last five years, he has tutored hundreds of successful Oxbridge and Medical applicants. He has also authored ten books on admissions tests and interviews.

Rohan has taught physiology to undergraduates and interviewed medical school applicants for Cambridge. He has published research on bone physiology and writes education articles for the Independent and Huffington Post. In his spare time, Rohan enjoys playing the piano and table tennis.

The Basics 6

2008 10
    Section 1 10
    Section 2 32

2009 37
    Section 1 37
    Section 2 57

2010 67
    Section 1 67
    Section 2 92

2011 100
    Section 1 100
    Section 2 124

2012 134
    Section 1 134
    Section 2 150

THE BASICS

What are TSA Past Papers?

Thousands of students take the TSA exam each year. These exam papers are then released online to help future students prepare for the exam. Before 2013, these papers were not publically available meaning that students had to rely on the specimen papers and other resources for practice. However, since their release in 2013, TSA past papers have become an invaluable resource in any student's preparation.

Where can I get TSA Past Papers?

This book does not include TSA past paper questions because it would be over 1,000 pages long! However, TSA past papers from 2008 are available for free from the official TSA website. To save you the hassle of downloading lots of files, we've put them all into one easy-to-access (and free!) folder for you at **www.uniadmissions.co.uk/tsa-past-papers**.

How should I use TSA Past Papers?

TSA Past papers are one the best ways to prepare for the TSA. Careful use of them can dramatically boost your scores in a short period of time. The way you use them will depend on your learning style and how much time you have until the exam date but in general, you should try to do at least 2008 – 2015 once. If time permits, do them twice- practice really does make perfect!

How should I prepare for the TSA?

Although this is a cliché, the best way to prepare for the exam is to start early – ideally by September at the latest for TSA Oxford and by October for TSA Cambridge. 4 weeks of preparation is usually sufficient for the majority of students. If you're organised, you can follow the schema below:

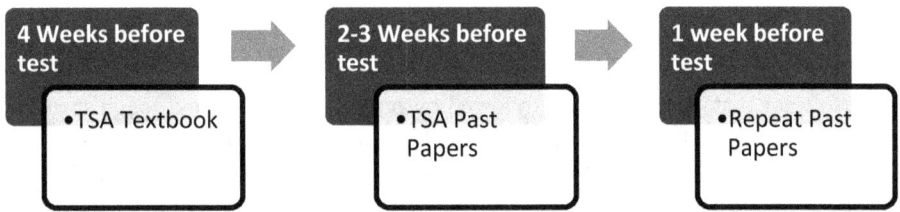

This paradigm allows you to focus your preparation and not 'waste' past papers. In general, aim to get a textbook that has lots of practice questions e.g. *The Ultimate TSA Guide* (**www.uniadmissions.co.uk/tsa-book**) – this allows you to rapidly identify any weaknesses that you might have e.g. identifying flaws, spatial awareness etc.

You are strongly advised to get a copy of '*The Ultimate TSA Guide*' which has 300 practice questions– you can get a free copy by following the instructions at the back of this book.

Finally, it's then time to move onto past papers. The number of TSA papers you can do will depend on the time you have available but you should try to do each paper at least once. If you have time, repeat each paper (choose a different essay question). Practice really does make perfect!

If you find that you've exhausted all past papers, there are an additional six mock papers available in *TSA Practice Papers* (flick to the back to get a free copy).

How should I use this book?

This book is designed to accelerate your learning from TSA past papers. Avoid the urge to have this book open alongside a past paper you're seeing for the first time. The TSA is difficult because of the intense time pressure it puts you under – the best way of replicating this is by doing past papers under strict exam conditions (no half measures!). Don't start out by doing past papers (see previous page) as this 'wastes' papers.

Once you've finished, take a break and then mark your answers. Then, review the questions that you got wrong followed by ones which you found tough/spent too much time on. This is the best way to learn and with practice, you should find yourself steadily improving. You should keep a track of your scores on the previous page so you can track your progress.

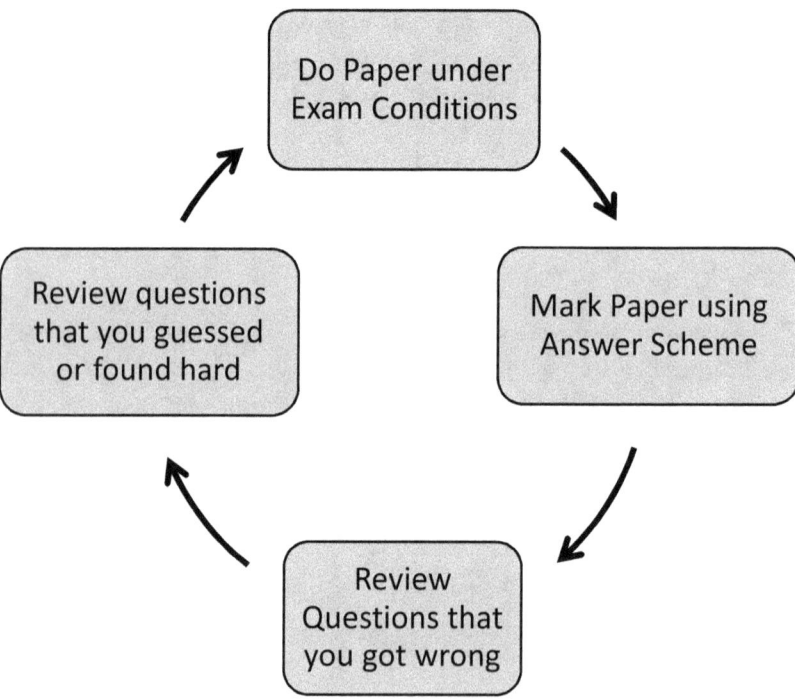

Scoring Tables

Use these to keep a record of your scores – you can then easily see which paper you should attempt next (always the one with the lowest score).

Volume One

Volume Two

Extra Practice

If you're blessed with a good memory, you might remember the answers to certain questions in the past papers – making it less useful to repeat them again. If you want to tackle extra mock papers which are fully up-to-date then check out *TSA Practice Papers* for **six** full mock papers with worked solutions (flick to the back for a free copy).

SECTION 1
Practice Pape

Practice Pape

2008

Section 1

Question 1: C
The passage discusses how measuring wealth relative to average income measures inequality, not poverty. It then goes on to describe a situation where under this flawed definition, a pay rise for some leads to some being described as in "poverty" (where this may not be the case), and how in some societies, there is such widespread poverty that using average income describes very few as being in poverty (though there are many in poverty).

Both A) and C) are valid conclusions from this passage, but we can see that the statement in A) goes on to support that in C). Thus A) is an intermediate conclusion, and C) is the main conclusion of this passage. D) Is a reason given in the passage to support these conclusions, and thus is not a conclusion in itself.
E) is an irrelevant statement, whilst B) is not a valid conclusion as the passage has made no reference to whether there is a suitable definition of poverty, it has simply refuted one possibility. This does *not* mean there are no others.

Question 2: C
Let a Child fare be C, and an Adult fare be A. We see from the question that 1 adult far plus 2 child fares is £1.20. Thus: $A+2C=120$(pence)
We also see that C is greater than 0.5A. Thus, 2C must be greater than A, so A must be less than half of 120. Thus, the answer cannot be D) or E), as A is at least half of 120 in both these answers.

We also see that C is less than A. Thus, if $A+2C=120$, 3 times A must be *more* than 120. Thus, A must be more than 40. This means the answer cannot be A) or B), because in these answers A is not more than 40. Thus, we can see that the answer must be C).

SECTION ONE 2008

Question 3: D
The passage discusses how the number of drug-related road deaths has increased more than drink-driving deaths in recent years, and concludes that this means the drink-driving campaigns have been successful, and it is now time to begin campaigns against drug-related driving.

Answer D) correctly points out that this reasoning is flawed. Just because Drug-driving has increased more does *not* mean the campaigns have been successful. We have no information on how much drink-driving deaths would have increased without the campaigns. This is the information we need to conclude that the campaigns have been successful. Observing that one problem is bigger does not mean another thing is not a problem. B) and E) are completely irrelevant statements, and thus are not flaws.

A) And C) both refer to the possibility that a campaign against driving on drugs may not be as successful as one against drunk-driving. These would weaken the argument, if true, but they are not *flaws* because they do not mean the argument's conclusions are invalid. A) and C) could be described as counter arguments, but are not logical flaws with the argument's reasoning.

Question 4: B
The passage discusses how socialist politicians are often criticised as being hypocritical for objecting to inequalities in wealth whilst enjoying above-average wealth. It then goes on to refute this criticism as invalid by claiming there is no hypocrisy in enjoying high levels of wealth whilst still arguing for a fairer society.
If all these reasons are true, they give us good cause to believe that one *therefore can* be a socialist (and thus argue for less inequality in wealth) whilst enjoying high levels of personal wealth. Thus, B) is the answer.

E) is the opposite of this conclusion, and directly goes against the thread of the argument, thus E) is not a valid conclusion from the passage.

Equally, the argument makes no claims of what socialists with wealth should do about those in even wealthier positions, so A) is incorrect and cannot be concluded from this argument.

Equally, the argument says nothing about whether wealth inequality is immoral, or who is most effective at fighting it. Thus, C) and D) are also incorrect.

Question 5: D

The passage discusses how the University has implemented improved safety features on areas of its campus, which will help to make those areas safer. It then describes how the council owns the lake, so the University has not been able to implement safety features in this area. It concludes on the basis of these reasons that the lake will be a dangerous area, and should be avoided by lone students at night.

However, this argument is flawed, as it assumes that the only thing making an area safer is the University's safety measures. The council could have put its own safety features in place, which may be just as good as the University's and this would make the conclusion invalid. Answer D) correctly illustrates this flaw.

Answers C) and E) are irrelevant. Whether students would pass the lake regularly, or the reasons behind the safety features, do not affect whether the area will be dangerous without the implementation of the safety features. Thus C) and E) do not affect the argument's conclusion.

Answer B) would actually strengthen the argument. If the council consider the lake as part of the campus it would suggest they have not implemented any safety features of their own, thus reinforcing the notion it may be dangerous.

Answer A) is not an assumption because it does not need to be true for the conclusion to be valid. Even if students do look after their own safety, the notion of avoiding walking around the lake alone at night might still be valid (in fact this may *be* a measure students could take to ensure their own safety).

Question 6: D

Since the rear wheel is 2.5m in circumference, it will complete a full rotation every 2.5m. Equally, the front wheel will complete a full rotation every 2m.
We know the tyre valves are beginning at the bottom of the wheels, so they will be in this position again after each complete rotation.

Thus, we are simply looking for the lowest common multiple of 2 and 2.5. This will give us the number of metres after which each wheel will have rotated a whole number of times, and will both be at the bottom at the same time.

The lowest common multiple of these numbers is 10. After this distance, the front wheel will have completed 5 rotations, and the rear wheel will have made 4 complete rotations.

SECTION ONE 2008

Question 7: E

The Question tells us that the last letter of each 5-digit code signifies which item this set of letters represents the code for, and that B represents the Bicycle lock combination. We see that the last 5-digit code ends with a B - D F G C B
Thus, D F G C represents the 4-digit code for the bicycle lock. Now we simply look at each letter and see which letter of the alphabet it is:
- D is the 4th letter
- F is the 6th letter
- G is the 7th letter
- C is the 3rd letter

Thus, we end up with 4673 as our combination. We are told that each 4-letter code is written in *reverse* with respect to the actual number code, so we know the bicycle lock combination is 3764.

Question 8: A

We can see in the top view a smaller square shape in the centre of the sculpture. This could be either a raised feature, going *out* of the sculpture, or an indent, going *into* the sculpture.

We can see from the top view that all the sides are flat, with nothing projecting out of them. However, we cannot see if any of the sides have any indented features, so we cannot judge if this might be the case.
A final thing we can see from the top view is that the edges of the top face, around the central feature on the top are all flat, with no indents or raised features.
View B) is possible, as it simply shows featureless sides, and a raised feature on top, which is possible.
View C) is possible because the top feature could be an indent, which would not be visible from the sides, and the sides could have indents represented by the square seen in view C).

View D) is possible as it shows a raised feature on top, which could be the feature in the top view, and a feature in the sides which could be an indent. This is possible from the information given in the top view.
E) is a possible view because it simply shows featureless sides, and the feature on the top may be an indent, which would not be visible from the sides.
A) is not a possible view because it shows indents along the top edge, other than the central feature. We can see from the top view that apart from the central feature, the top is all featureless, with no indents. Thus A) is not possible.

Question 9: C
The passage discusses how in evolution, organs do not develop to be large unless they are used. Based on this fact, the passage describes how whale brains are much larger than our own, and concludes that sperm whales must therefore possess great intelligence, perhaps beyond our own understanding.

Answers A) and B) are irrelevant. Language does not necessarily equal intelligence (so B) is wrong), and the fact that intelligence is possible without large brains does not mean that large brains do not give intelligence (as implied in the passage).

D) and E) actually strengthen the argument. D), by saying the large brains are not related to the large size of sperm whales, reinforces the idea they might be for intelligence, whilst E) reinforces the notion their intelligence may be beyond our understanding.

C) however, weakens the argument, because if the brains may be used for something else then their size does not necessarily mean they are used for intelligence, thus weakening the argument.

Question 10: C
The passage can be summarised very simply. It argues that Mrs Jackson **will** resign if Mrs List is promoted, and then argues that if Mrs List is **not** promoted, Mrs Jackson will **not** resign.

Answer C) directly contradicts how this reason leads on to this conclusion. It points out that the conclusion is invalid because it does not follow on that Mrs Jackson **won't** resign if Mrs List is not promoted. Thus, C) correctly identifies a flaw in the argument.

A), B) and E) are completely irrelevant because they have no effect on the fact that Mrs Jackson will resign if Mrs List is promoted, or whether this means she will not resign if Mrs List is not promoted. Thus, they are not flaws with the argument.

D) is not a flaw because the argument directly states that Mrs Jackson **will** resign if Mrs List is promoted. Therefore, we assume there is no a chance that she won't, as it has been directly stated in the argument that she will.

SECTION ONE 2008

Question 11: D
The passage discusses how there is evidence that Cannabis has therapeutic uses in certain medical scenarios, and that it should thus be legal. It goes on to say this means doctors being allowed to prescribe cannabis would help those with the relevant medical conditions. It then goes on to say legalising cannabis prescriptions would also allow large-scale studies to establish if the supposed benefit is real.

If we accept all of these reasons as true, we have good reason to believe doctors should be allowed to prescribe cannabis. Thus, D) is the main conclusion of the passage.

B) is a reason given in this argument which supports the main conclusion.

A) and C) are not conclusions or reasons given in the passage. Both could be said to strengthen the conclusion, if true, by providing further reasons why current laws should be changed. However, neither can be concluded from the passage, and are thus irrelevant to what the main conclusion is.

E) is an irrelevant statement with no effect on the argument's reasoning or its conclusion.

Question 12: C
Let the normal rate be termed Y, and the overtime rate be termed Z.
We know from the Question that Pierre worked 30 sessions at normal rate, and 10 at overtime rate, and earned 700 euros. Thus, $30Y + 10Z = 700$
Equally, we know that Mark worked 20 sessions at the normal rate, and 5 at overtime rate, and earned 425 euros. Thus, $20Y + 5Z = 425$
By subtracting the 2nd equation from the first, we can see that:
$30Y - 20Y + 10Z - 5Z = 700 - 425$
$10Y + 5Z = 275$
We can then subtract this new equation from the equation formed from Mark's work, to work out the normal rate. Thus:
$20Y - 10Y + 5Z - 5Z = 425 - 275$
$10Y = 150.$ Thus $Y = 15$
Now that we know Y, we can now use either waiter's working patter and pay to calculate z.
Let us use Pierre's work and pay. We see that 30Y+10Z=700. Thus, $30 times 15 + 10Z = 700$
$30\ times\ 15\ is\ 450.$ Thus, $450 + 10Z = 700$
Thus $10Z = 700 - 450$, which is 250.
If 10Z=250, Z must equal 25.
Thus, the overtime rate is 25.

SECTION ONE   2008

Question 13: E
This question is simpler than it appears on first glance.
We are calculating the number of wins, so we need to combine the away wins and home wins into one figure. Since they are next to each other on the pie chart we can easily do this visually.
Now, if we examine the amount of the chart made up of "wins", and the amount of the chart made up of "draws" (both home and away combined), we can see that roughly ¾ of the pie chart is made up of some form of "wins". We can verify this by examining the angle between the boundary of "Home Wins" and "Away Draws", and the boundary between "Away wins" and "Home Draws". This angle appears to be 90 degrees. This verifies that ¾ of the pie chart is made up of wins, whilst the other quarter is made up of draws.
Three quarters of 24 is 18.

Question 14: E
We can readily see that when this net is folded up, the face with the cross will be adjacent to the 4 faces with thick grey lines, and that the thin diagonal line will be on the opposite side. Most of the incorrect answers can be discounted in terms of what faces will be adjacent to each other:
Shapes A) and B) cannot be made because the 2 non-diagonal thick grey lines *must* be located on opposite sides of the box (separated by the cross as seen in the Net) and cannot be adjacent.
Shape C) cannot be made because the face with the thin diagonal line *must* be opposite the face with the cross, and cannot be adjacent.
Shape D) and Shape E) both present faces adjacent to each other in a possible combination, so we must now examine the orientations these faces will be to each other when folded

Shape D) is not possible because the orientations of the diagonal lines are incorrect. We can see that there are definitely possible combinations in which one of the thick grey diagonal line will be orientated with one of the vertical or horizontal thick grey lines as shown in view D). However, we see that in order for this to be achieved, the thin diagonal line must be *perpendicular* to the thick grey diagonal line from the viewpoint in question, *not* parallel as shown in view D).
Shape E) is possible. We can readily see if we turn the net upside down that we now have the Cross, with one thick grey line continuing upwards from the top of the cross, and one thick grey diagonal line radiating away from the bottom-right corner of the cross. This is exactly the view shown in E). Thus, E) is possible. Hence E) is the answer.

Question 15: D

The passage describes how random drug testing in prisons has caused prisoners to switch from Cannabis to Heroin, as it stays in the system for much less time. It describes how since drug testing was introduced, cannabis use has decreases but heroin use has increased. From this, we can conclude that drug testing **hasn't** solved the issue of drug use in prisons, because heroin use has increased. Thus D) is a valid conclusion from the passage.

The fact that it hasn't solved the issue does not mean that it **can't** solve the issue, so C) is not a valid conclusion. Equally, we cannot conclude that the method needs to be improved, so A) is incorrect.

The other 2 answers concern incorrect statistical inference. The fact that cannabis has increased and heroin has decreased does not mean heroin use is now higher, so B) is incorrect. Equally, the fact that heroin encourages intimidation doesn't mean that doubling heroin use will increase intimidation by the same amount (we have no information on *how much* heroin increases intimidation).

Question 16: D

The passage concludes that doctors must deceive their patients in some medical scenarios. The reasons given to support this are that patients have a right to know the truth, but in some scenarios the patient's health would be better served by not knowing the truth.

We can see that this reasoning only supports this conclusion if we accept the statement in D) as true. At no point has D) been stated, so it is therefore an assumption on which the argument rests.

E) is the opposite idea to the assumption in D), and if true, E) would invalidate the argument. Thus E) is not an assumption.

C) is irrelevant as it refers to a situation where a patient would be upset, and not where their health would worsen from knowing the truth. B) is also irrelevant in a similar way, referring to a situation where patients would be frightened, not risking their health.

A) is completely irrelevant, as what patients accept about a doctor's responsibilities does not affect how doctors should respect whether their right to know the truth is more important than their health.

SECTION ONE — 2008

Question 17: A
The passage concludes that tits that tear paper are searching for food, on the basis that they use a similar method to tear paper as they do to strip tree bark and search for food.
However, if we accept A) as true, then it seems the tits are choosing to tear paper *instead* of taking food which is readily visible and available, weakening the conclusion that tearing paper *is* a search for food.
B) is not correct because the fact that humans know does not mean that the tits themselves know that no food is forthcoming.
C) is incorrect because the fact that families feed them does not mean that the birds will not search for more food.
D) is an irrelevant statement, and thus incorrect.
E) is not correct because the fact that animals engage in pointless activities does not necessarily mean that this activity is pointless and not functional.

Question 18: E
With a rate of inflation of 10%, a car worth 500 Bols a year ago will now be worth 550 Bols (we assume for this question that the car's value has not degraded due to damage or wear and tear).
80% of 550 Bols is 440 Bols. Thus he will receive 440 Bols for his old car.
The new car will be worth 550 Bols, as it is an identical model, and with an inflation rate of 10%, something worth 500 Bols a year ago will now be worth 550 Bols.
550-440=110. Thus, Evitan will need to contribute 110 Bols towards purchasing the new vehicle.

Question 19: B
Answering this question first requires understanding how to read the table.
The left-hand side of the table describes the percentage *change* of industrial growth for the last year. Thus, we only need to consider the Latest figure, as this gives us the percentage change that the question refers to.
The right-hand side of the table describes the percentage *of the population* unemployed. Thus, to calculate the *change* in percentage unemployed we need to subtract the current figure from the figure a year ago.
Working these 2 figures out for the UK we see that it had a percentage increase of industrial production of 4.2%, as seen in the table, and that the % of the population unemployed fell from 10 to 8.8, a fall of 1.2.
Thus, we are seeking a country with a percentage increase in industrial production of >4.2, and a percentage *fall* in unemployed of more than 1.2 (i.e. a percentage change of less than minus 1.2).

SECTION ONE   2008

We see that all countries except the Netherlands and Belgium had a percentage change in industrial production of more than 4.2. However, of these countries, only Canada *also* had a fall in the % of people unemployed of more than 1.2

Question 20: B
First we should pick out the distances mentioned for the towns the two drivers have travelled to, in order to get an idea of where the towns are with respect to each other.
- Barneyville is 5km West, and 15km South of Abbeytown
- Carloston is 5km East of Barneyville.
- Denburgh is 5km South of Barneyville.
- Easterby is 10km East and 15km North of Denburgh.

We know that Ahmed is in Carloston, and Wayne is in Easterby. From the distances mentioned above, we can work out how far away Carloston is with respect to Easterby:
- Carloston is 5km East of Barneyville.
- Easterby is 10km East and 10km North of Barneyville (Denburgh is 5km south of Barneyville, and 0km East or West. Easterby is 10km East and 15km North of Denburgh)
- Thus, Easterby must be 5km East and 10km North of Carloston.

We know that all roads run directly North-South or East-west, so in order to get between these two towns, one driver must drive a 5km stretch, and then a 10km stretch. Thus, a total of 15km must be driven.

Question 21: D
The passage discusses how a study has found that those with more control over their work have lower levels of a particular stress-related disease. It then concludes that in order to reduce stress-related disease, we must give workers more control over their work.

The argument has confused cause and correlation twice. It has assumed that a lower incidence of heart disease is *because* of a lower incidence of stress, and not simply correlated with it. Answer D) correctly illustrates this flaw.

Answer A) is not a flaw because the argument does not imply the government is serious, it simply comments on what must happen if it is.

B) is not a flaw because whether workers wish to have more control is irrelevant to whether more control over work reduces stress-related disease.

C) is irrelevant as practical limitations of a policy do not affect what results this police will have *if* it is implemented.

E) is also irrelevant. Reducing the amount of stress-related disease among white-collar workers would still reduce the overall incidence. Thus, whether other groups are also affected is irrelevant.

SECTION ONE   2008

Question 22: B

The passage can be summarised as arguing that "A can happen if B happens. B has not happened, so A cannot happen". Although this is incorrect, we are still able to assess which of the answers follow this pattern.
- ➢ Answer B) follows this pattern, where "A" is curing a headache, and "B" is being willing to try acupuncture.
- ➢ Answers A) and C) both follow a pattern of "A can happen if B does. B happens, therefore A will happen". (In C), "A" is avoiding being overweight)
- ➢ D) can be summarised as "A happens if B does. A already happens so we do not need B".
- ➢ E) can be summarised as "A happens if B happens. A is happening, so B must be happening".

None of these are the same as the reasoning used in the passage.

Question 23: C

The passage refers to a principle where *actions* bring more responsibility for a given scenario than a *failure to act*. This is because when an action is taken, the result would not have happened without the action from the person in question. However, when an action *fails* to be taken, the consequences would have happened anyway without the existence of the person who failed to act.
- ➢ Answer C) directly follows this principle, stating that those who break the law are more responsible than those who fail to prevent them from doing so.
- ➢ Answer A) follows an opposite principle, stating that the government is responsible for high crime rates through its *failure to act*, rather than through any given action.
- ➢ Answer B) simply states that a situation is happening, so clearly prisons are failing to prevent this situation. This does not give any claims on whether the prisons are therefore *responsible* for the situation.
- ➢ Answers D) and E) are irrelevant scenarios, which bear no relevance to the principle.

Question 24: D

First we need to work out how many performances will be carried out in the course of 1 full week. We see that there are 6 evening performances (one for each day except Sunday), and that in addition to this there are 3 Matinee performances (Wednesday, Thursday and Saturday). This gives a total of 9 performances in the course of each full week.

Now we need to work out how many complete weeks there are in each month. 7 goes into 31 fully 4 times, with 3 left over. Thus, there will be 4 full weeks each month. This gives a total of 36 performances.

SECTION ONE   2008

Then, there are 3 leftover days at the end of each month, which will contain extra shows. The maximum number of shows which occur in a 3 day period is either the period from Wednesday to Friday, or the period from Thursday to Saturday. In each of these 3 day periods there are a total of 5 performances.

Thus, the *maximum* number of performances that can occur in a month is 41.

Question 25: D

To calculate the number of days of labour that peter must pay for, we simply calculate the area of the filled in grey blocks, which represent a number of days of labour, and how many individual workers need to be present for those days. We can see that for all tasks except brickwork, only one row is filled in, so only one worker needs to be present.

➤ Preparing foundation lasts for 6 days, and one labourer, so accounts for 6 days labour in pay.
➤ Brickwork lasts for 9 days, and involves 2 labourers, so accounts for 18 days of labour in pay.
➤ Roofing lasts for 16 days, and involves 1 labourer, so accounts for 16 days of labour in pay.
➤ Glasswork lasts for 10 days, and involves 1 labourer, so accounts for 10 days of labour in pay.
➤ Decorating lasts for 8 days, and involves 1 labourer1, so accounts for 8 days of labour in pay.

This gives a total of 58 days of labour that must be paid for.

Now we examine how many is the maximum number of workers that may be on site at any one time. We can quickly see that there are 2 points when 3 roles are being carried out at the same time (Days 12-13 and Days 20-21)

One of these sets of days (Days 12-13) involves brickwork and 2 other jobs. We have already seen that brickwork requires two labourers, whilst all other jobs require 1 labourer. Thus, at this point there will be 4 labourers on site at 1 time.

Thus the answer is D), 58 and 4.

Question 26: B

This question holds a simple answer, and is designed to test your ability to think logically and clearly under time pressures.

The tile shown has 1 fully rounded curve, and shape made out of straight lines, which looks like a square with part of the corner cut off.

We can immediately see from examining the possible answers that Answer B) shows four tiles (Bottom-left, Middle-left, Bottom-Middle and the Middle tile) which only have rounded curves. Thus, this pattern cannot be made with tiles identical to that shown in the Question.

Question 27: C

The passage discusses how market forces have always limited the adverse effects of the environment on food production. However, it then goes on to say that we are not changing the environment irreversibly, with changes that will be permanent. The passage says how this will cause devastating effects on world agriculture, such that market forces alone cannot prevent a global famine.

Thus, the argument mentions Answers A), B), D) and E) as reasons throughout the passage.

The argument then brings all these reasons together and concludes that some form of planned intervention is necessary. Thus, we can see how answers A), B), D) and E) all come together to support the statement in C). Thus, C) is the main conclusion of this passage, and all the other answers are reasons given to support this conclusion.

Question 28: A

The passage discusses plants being genetically modified to produce pesticides. However, it then discusses how pesticides lose their effectiveness if used continually, thanks to development of resistance in the pests, and that they remain effective only if there are gaps in which they are not used. It then finishes off by saying that the plants which produce pesticide will produce them continuously, without these gaps.

If all these factors are true, we have good cause to believe that the Pesticides produced by these plants will become ineffective against pests. Thus A) is the answer.

Nothing in the passage says anything about plants becoming contaminated with pesticides, so B) is not a conclusion.

All the other answers actually argue *weaken* the conclusion given in A), or contradict stated reasons in the passage. Thus, none of these are valid conclusions from the passage.

SECTION ONE 2008

Question 29: D
The passage states that recycling is not the best answer to the problem of garbage disposal. It backs this up by describing how most bottles in America are constructed of PET. It then discusses how this material is not safe when recycled, due to the absorbance of other chemicals which may then leach into food and drink stored in the recycled product.
We can see that this conclusion that recycling is not the best solution relies on an assumption that there is little use for recycled PET other than food/drink containers. If there are other, safe uses, then the potential carcinogenicity described is no longer a reason not to recycle. Since the argument has provided no other reasons why recycling is not the best solution, its conclusion is no longer valid from its reasoning. Thus, D) correctly identifies an assumption in the passage.
The passage says nothing about the pros/cons of burning plastic, reducing plastic production, the exhaustion of landfill space or whether soft drink bottles should be made of other material. Thus, none of the other answers affect the argument's conclusion, and thus they are not assumptions.

Question 30: B
We can see from the information given that 8 bricks are required to cover an area of 40cm by 40cm.
40cm by 40cm is 1600cm^2.
The area of the whole driveway is 550cm by 400cm. This is 220,000cm^2
220,000cm^2 divided by 1600cm^2 is 137.5 (140 times 1600 is 224,000. 2.5 times 1600 is 4000).
Thus, the amount of bricks required will be 8 times 137.5. 140 times 8 is 1120. 2.5 times 8 is 20.
Thus, 1100 bricks will be needed to cover the entire driveway.

Question 31: B
The ratio is currently 1/3 Sand to 2/3 Coir. Thus, there is currently 5kg of Sand present in the mixture, and 10kg of Coir.
The target ratio is 3/5 sand to 2/5 Coir. Since we are only adding sand, the actual amount of coir will stay the same. Thus, 2/5 of the entire weight of the new compost will be 10kg.
If 10kg is 2/5 of the entire weight of the new compost, then the full weight must be 25kg. Since 3/5 is Sand, this means there needs to be 15kg of Sand to make the new mixture.
Thus, 15kg of sand is needed, and 5kg is present. Thus, 10kg of Sand must be added to make the new compost. Hence, the answer is B).

SECTION ONE 2008

Question 32: E
If the second stage begins with 4 quarter finals then the number of teams progressing through to the second stage must be 8, as there will be 2 teams in each of the 4 quarter finals. To progress through to the next round, you must have won your group, so there must be 8 group winners and hence 8 groups. The teams are divided into equal groups, so the number of teams starting the competition must be a multiple of 8. The two multiples of 8 given as possible answers are 8 and 72. If the competition started with 8 teams then there would be 1 in each group and hence no need for the first stage at all, so this cannot be the answer. Hence the only possible solution of the 5 given is 72. Hence the solution is E.

Question 33: C
The passage argues catalytic converters were created to tackle one kind of pollution, but in fact create another. It argues that converters do this by removing certain pollutants that usually serve to reduce the levels of ground-level ozone. The passage claims that since such ozone is a major cause of choking smogs, people with breathing difficulties affected by it will be worse off if more catalytic converters are used.

We can see that this conclusion (that catalytic converters indirectly cause breathing difficulties) is weakened by the statement in C), which suggests that catalytic converters remove other pollutants that also cause breathing difficulties. If this is the case, it's overall effect on breathing difficulties may be neutral or a positive one (if the pollutants it removes are more potent than ground-level ozone). Thus C) weakens the argument, and the answer is B).

A), B), D) and E) are all irrelevant in that they do not affect whether increasing the number of catalytic converters in cars will increase the incidence of breathing difficulties, via indirectly increasing the amount of ground-level ozone.

SECTION ONE   2008

Question 34: A
The passage discusses how membership of the EU *has resulted* in growth and stability in the EU members. It then discusses how it is desirable for previous Eastern Bloc members to become more prosperous and stable, and concludes that they should therefore join the EU.

Answer A) points out that this conclusion relies on the former Eastern Bloc countries having the same potential as the current EU members. If they do not, there is no guarantee that membership of the EU would have the same effect, and thus the argument's conclusion is invalid. Hence, the answer is A).

Answers B) and D) appear to be flaws, but are not upon closer inspection. Answer B) relates to other non EU members also being prosperous and stable, whilst D) refers to exclusion from the EU not necessarily resulting in lack of prosperity/stability. Neither of these statements necessarily means that EU membership would not bring prosperity and stability, and thus they do not affect how the argument's reasons support its conclusion.

C) and E) are completely irrelevant statements which have no effect on whether EU membership brings prosperity and/or stability. Thus they are not flaws.

Question 35: A
The passage argues that the driver's attempt to disable his main competitor and thus ensure victory was fair. It argues that the other driver would have done the same, that the driver was acting as anyone else would, and that he was defending an earned lead. All of these reasons, if true, give us good cause to believe that the main driver's actions were not unfair. Thus, A) correctly identifies the main conclusion of this passage.

B) is an irrelevant statement, as the rewards have no effect on whether the driver's actions were fair, and the fact that there are rewards is simply a stated fact, and not a conclusion.

C) is not a valid conclusion from the passage, and in fact disagrees with the passage, as the passage claims the actions "may have been dangerous and irresponsible". Thus C) is not a conclusion to the argument, and appears to be a counter-argument to the passage.

D) and E) are both reasons given in the passage to support the main conclusion, which is that given in Answer A).

SECTION ONE   2008

Question 36: B
Firstly we need to calculate what time the first flight arrives in Dubai, as this is not given. We are told that the wait time in Dubai is 5 hours and 15 minutes, and the flight leaves Dubai at 14:30 on Saturday. Hence the flight from London must arrive at 09:15 Dubai time on Saturday (5 hours and 15 minutes before 14:30).

Now we can work out the time spent in the air for each flight. The first flight departs London at 22:30 on Friday and arrives in Dubai at 09:15 on Saturday, both given in local times. If Dubai is 4 hours ahead of London, then 09:15 in Dubai will be 05:15 in London. Hence the flight is in the air from 22:30 until 05:15 the next morning, London time. This is 6 hours 45 minutes.

The second flight departs Dubai at 14:30 on Saturday and arrives in Kampala at 20:45 on Saturday, both given in local times. If Kampala is 1 hour behind Dubai then 20:45 in Kampala will be 21:45 in Dubai. Hence the flight is in the air from 14:30 until 21:45, Dubai time. This is 7 hours and 15 minutes. The total time spent in the air is hence 6 hours 45 minutes + 7 hours 15 minutes, which is 14 hours. Hence the solution is B.

Question 37: B
If each pen must be accessible from at least one side, the most efficient way to pen the sheep in is to have them in rows of 2, as this means most of the hurdles are being used to pen in more than one sheep. If the sheep are in 2 rows of 8 then 3 lots of 8 hurdles will be needed to pen them in one direction, and 9 lots of 2 will be needed in the other direction as shown below.

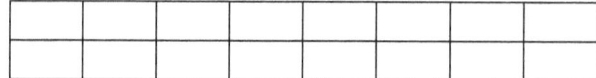

Hence the minimum number of hurdles required is 3x8 + 2x9 = 24 + 18 = 42. Hence the solution is B.

Question 38: C
The answer to this question is best illustrated by a diagram:
This is a representation of the ribbon folded in half:

If this is then folded in half again, it will look somewhat like this:

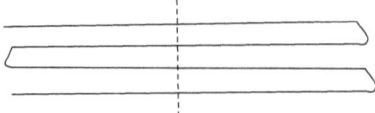

If we cut through the middle of here as shown by the dotted line, and consider the pieces we will get starting from the top left and following the ribbon, there will be 1 of 1m, 1 of 2m, 1 of 2m, 1 of 2m, then 1 of 1m. Hence there will be 3 pieces of 2m and 2 of 1m. Hence the answer is C.

Question 39: E
The passage discusses how ecotourism *should* provide a sustainable alternative to the overuse of natural resources, but then describes how it actually causes a range of problems in wildlife, such as new diseases, stress and reduced breeding success. From this information, we can infer that whatever benefits ecotourism provides are outweighed by these negative consequences. This is especially true because the argument implies that ecotourism *should* provide a sustainable alternative, which carries an implication that in truth it doesn't. Thus E) can be reliable inferred.
We cannot imply anything about the nature of the changes observes such as new diseases and altered behaviour patterns, so the suggestion in A) that they may not be apparent to a casual observer *cannot* reliably be concluded from the passage.
C) and D) are irrelevant statements on which we have no information (e.g. we do not know if the stressed dolphins become frenetic when near fishing boats), thus neither of these can be reliably inferred.
Equally, the passage is actually implying that ecotourism projects *are not* sustainable thanks to the harm caused to wildlife. Thus, B) cannot be reliably inferred.

Question 40: A
The passage claims that his lack of care about winning *fully explain* why he did not win the open championship.
We can readily see that this conclusion rests entirely on an assumption that if he had cared more about winning, he would have won. If this is not true, then this conclusion that the lack of care*fully explains* the loss is not valid. Thus A) correctly identifies an assumption in the passage.
B) and C) are both invalid conclusions from the passage. The idea that lack of care was all that stopped Van de Velde winning is *not* enough to assume that caring enough is *sufficient* to win major championships. There are probably many other things required, which Van de Velde possessed (such as a good level of skill). Equally, we cannot state that there *is* nothing worse than losing from the suggestion that this belief needs to be held in order to win championships.
E) is also an invalid conclusion from the passage. The idea that Van de Velde would have won the championship if he'd cared more about winning does suggest that he should have cared more about losing **if** he wants to win the championship.
D) is an irrelevant statement for which the passage provides no information.
Thus, only A) is an assumption in this passage, and the answer is A).

SECTION ONE   2008

Question 41: B

The passage describes how extended warranties for electrical items are poor value for money, because the average cost of repair per customer is less than the price of the warranty. It then concludes that customers would be well advised to not purchase the extended warranties due to the poor value for money.

Answers A), C) D) and E) all strengthen this conclusion by reinforcing the notion that the warranties are poor value for money.

Answer B) is the only answer which weakens this conclusion. If the actual cost for many people is 0, then it may be that the average cost of repairs *if a repair is needed* is greater than the cost of the warranty. Thus, the warranties may actually be *good* value for money if a repair is required. Thus, B) weakens the argument.

Question 42: C

- For a parcel which requires 24c postage, they will require 4 6c stamps.
- For a parcel which requires 30c postage, they will require 1 30c stamp
- For a parcel which requires 72c postage, they will require 2 6c stamps and 2 30c stamps

(Note: the 30c and 72c postages could be made using other combinations but it is specified in the question that we use the minimum number of stamps on each parcel.

The parcels are sent in about equal numbers, so the proportion of stamps needed will be the proportion that would be needed if 1 parcel of each postage is sent.

If 1 parcel which requires each amount of postage is sent, then 6 6c stamps and 3 30c stamps will be required (adding up the amounts given above). Hence 2 6c stamps are needed for every 1 30c stamp. Hence the solution is C.

Question 43: C

The Goats came 3rd in League B, so they will play in the 4th quarterfinal and hence, if they are to progress to the final, the 2nd semi-final. Hence their opponents in the final must come from the 1st semi-final. In the first semi-final, the winners of the 1st quarterfinal play the winners of the 2nd quarterfinal; hence the teams that could potentially play The Goats in the final are those who play in the first two quarterfinals. This is the winners of League A (The Scorpions), 4th place in League B (The Archers), 3rd place in League A (The Bearers) or the runners up in League B (The Bulls). Hence the teams The Goats may face in the final are The Scorpions, The Archers, The Bearers and The Bulls. Hence the solution is C.

Question 44: E
One piece of carpet is 4m x 6m and the other is 4m x 4m. The total amount of carpet is 40 square metres. We can rule each of the other 4 rooms out in turn.

The Snooker Room is 8m by 6m, 48 square metres, so there is not physically enough carpet to use it in this room, as there would be a 4m x 2m gap somewhere.

The Breakfast Room is 7m by 5m, 35 square metres. Although there is physically enough carpet for this room, carpeting this room with it would require more than one join as whichever way the larger piece of carpet is laid, the smaller piece cannot cover the space that is left without being cut into two pieces.

The Dining Room is 6m by 6m, 36 square metres. Although there is physically enough carpet for this room, carpeting this room with it would require more than one join as whichever way the larger piece of carpet is laid, the smaller piece cannot cover the space that is left without being cut into two pieces.

The Living Room is 8m by 5m, 40 square metres. Although there is physically enough carpet for this room, carpeting this room with it would require more than one join as whichever way the larger piece of carpet is laid, the smaller piece cannot cover the space that is left without being cut into two pieces.

The Library is 9m by 4m, 36 square metres. We can use the carpet for this room because by laying the carpet end to end we can cover both the length and the width of the room with the carpet available, with only one join.

Question 45: A
The passage discusses how a number of farm animals have been found killed in fields, and that the injuries suggest a large predator is responsible, which locals claim is some form of big cat. The passage then discusses how official investigations have refuted these claims, and say that the sightings were of big cats. However, the argument claims the investigations are flawed in that although the sightings may be explained by domestic cats, the injuries cannot be, and concludes that big cats such as pumas must therefore be responsible.

This last statement is supported if we accept all the other reasons in the argument as being true, and thus this is the main conclusion of the passage. Thus A) correctly identifies the main conclusion of the passage.

C) is a reason given in the passage to support this main conclusion, and is not a conclusion in itself.

D) and E) are irrelevant statements which do not affect the argument's conclusion, and are not conclusions in themselves.

B), meanwhile, is a direct contradiction of the results of the investigations, which is conceded as being true by the passage. Thus, it is not in any way a conclusion from the passage.

Question 46: A

The passage can be summarised as claiming that "A" and "B" are needed for "C". Although we have "B", we do not have "A", so we cannot get/do "C".

Only answer A) follows this same structure. Here "A" is having enough time to get to the station, "B" is having enough time to find the platform, and "C" is catching the train.

Answer B) refers to a situation where "A" and "B" are both needed for "C", but in instance *neither* of "A" and "B" has been provided, so "C" cannot happen. This is different from *just one* of "A" and "B" being missing, so "C" cannot happen.

Answer E) refers to a situation where *either* "A" or "B" is sufficient for "C", but neither have been provided, so "C" can't happen. Again this is different from the question.

Answer C) refers to a situation where only one criterion needs to be fulfilled (i.e. the tide holding back for long enough for the truck to arrive), but it cannot be. This is different from needing two criteria, as in the question.

Answer D) refers to a completely different setting where one of two options needs to be used to carry out "C", and one of the options cannot be used, so the other must be.

Hence, the answer is A)

Question 47: B

The passage illustrates a principle where only those who use a certain service should have to pay towards its upkeep, and how nobody who does not use the service should have to pay for it.

Answer B) follows this same principle, suggesting that those who do not visit London should not have to pay towards the subsidisation of the London transport system.

Answers C) and D) both refer to a principle where all people should provide a contribution to services which are beneficial, even if they do not use them themselves. Thus, these two answers illustrate an opposite principle to that in the question.

Answers A) and E) do not really refer to a principle at all, and simply describe possible outcomes from not subsidising art galleries. These answers are probably provided to distract you, and test you ability to focus on the *principle* at hand (whether people should pay towards a service they don't use) rather than the *situation* at hand (art galleries).

SECTION ONE   2008

Question 48: C
- The winner receives $100 for lifting 80kg.
- He then receives another $10 for having lifted 85kg.
- He then receives another $15 for having lifted 90kg.
- He then receives another $20 for having lifted 95kg.
- He then receives another $25 for having lifted 100kg.
- He then receives another $30 for having lifted 105kg.
- He then receives another $35 for having lifted 110kg.
- Hence his total prize money is $100 + $10 + $15 + $20 + $25 + $30 + $35 = $235.
- Hence the solution is C.

Question 49: D
Energy costs can be considered Gas + Electricity + Coal + Logs. We can add the costs of these up for each month and this elicits:

January: 125	May: 180	September: 0
February: 160	June: 80	October: 70
March: 40	July: 80	November: 0
April: 120	August: 80	December: 85

The highest of these is May, at 180. Hence the solution is D.

Question 50: D
The only boxes on the right hand side as Vinod looks at the boxes are directly in front of Sarah, so this pile of boxes must be 3 high. One of the piles in the middle as Vinod looks at the boxes must be 4 high but none can be more than 4 high, and one of the piles on the left as Vinod looks at the boxes must be 2 high but none can be more than 2 high.

We have already ascertained from the view than Vinod has that the pile of boxes in front of Sarah must be 3 high, so the middle pile as Sarah looks at the boxes must be at least 3 high. Hence D cannot be the view that Sarah sees as the middle pile is only 2 high. Hence the solution is D.

END OF SECTION

Section 2

When, if ever, is forgiveness wrong?
This is a complex philosophical question based on an empirical issue. In order to answer this question, which I would not advise unless you have a solid and wide philosophical background, it is vital that one has a sound understanding of the different ways that the term 'wrong' can be interpreted. Explore and define this term in as clear a manner as possible (some pointers are given below). Failure to either define this word concretely and clearly, or to stray from the definition introduced in the opening paragraph will result in an essay which is vague and weak in terms of argument – the worst possible characteristics of a philosophy essay.

Introduction
- Definitions - to write a good essay, it is vital that two words are clearly defined, and that the definitions are followed throughout. Those words are 'forgiveness' and 'wrong'
- A good idea would be to use a case (if possible from real life) to illustrate your argument – such as, for example, a murder.
- Lay out your argument and give a good idea of the direction you are going to take.

Forgiveness
- Consider a clear definition of forgiveness – for example 'to stop feeling angry or resentful towards (someone) for an offence, flaw, or mistake'.
- A good idea would be to explore what forgiveness is in terms of victim and perpetrator– namely that the perpetrator commits an act which in some way harms the victim, and forgiveness is a decision by the victim to accept what had happened.
- It might also be a good idea to contrast forgiveness with reconciliation – reconciliation goes beyond forgiveness in that the forgiveness is the acceptance of a bad act, whereas reconciliation is the restoration of good relations.
- A key element of forgiveness, which should be explored is the idea of repentance on the part of the perpetrator.

Wrong
➤ Consider a clear definition of wrong. Explore the idea of wrongness in different senses – for example in the context of justice and the law, or a more personal interpretation.

Possible arguments
➤ One possible approach is to flip the question, from 'when is forgiveness wrong' to 'when is forgiveness right?' A way to answer could be to refer back to your example and think of instances where forgiveness is right – such as for example when the perpetrator has shown remorse.
➤ The idea of remorse serves as a good focal point to build your argument around – namely forgiveness can only be 'right' when the perpetrator shows remorse.
➤ A counter argument to this could be that certain crimes are unforgivable (give examples), no matter how much remorse is shown

Conclusion
➤ Summarize the main points made on each side of the argument in the essay.
➤ Link back to the original definitions you gave – is there any way that the definitions could be changed, and, if so, would that lead to a different conclusion? For example you may come to one conclusion if you focus on a moral definition of the word wrong, but perhaps, if you take a more legalistic definition of the word you may come to the conclusion which is the opposite of what you had previously. This is a useful perspective to introduce when you conclude your essay as it shows wider thinking around the topic.
➤ A key thing to keep in mind is that this is a humanities and not a science question meaning that there is not a definite answer.

SECTION TWO — 2008

Should parking fines be based on the driver's income?

This essay is relatively simple to answer if structured correctly due to the availability of arguments both for and against the point. However, it is important that the arguments you use are measured and focused in approach – merely writing down every single argument you can think of will not lead to a good essay as you will fail to achieve sufficient depth or clarity of thought.

It is therefore very important with this question to have a clear and concise opening paragraph in which you introduce the reader to the arguments that you are about to make. As well as framing the essay well, this has the additional benefit of helping you keep the rest of the essay structured.

Introduction
- Identify the key concepts in the question – namely justice/punishment and fines - and explain them in relation to the question. In this case parking fines are in place to prevent a negative action from taking place (bad parking) and the level of the fine – the key point of the question – determines the effectiveness of the fine.
- If parking fines were £10,000, people would be very careful to park correctly. However, there is another element to this question – justice. Parking incorrectly is a problem, but it is important that the punishment reflects the severity of the crime. This is why parking fines are limited.
- Another thing to consider and explore is relative income - £10,000 means a lot more to someone earning £20,000 a year than to someone who earns £2 million.

Possible arguments in favour
Equality of impact of law
- This is the idea of relative income. The purpose of a fine is to ensure that the offender faces the consequences of their actions. The extent to which a financial penalty feels like a negative consequence is relative to someone's income; not to the amount that the fine is.
- Therefore, if you make fines proportional to the income someone has, everybody feels the impact of the punishment equally, rather than the poor facing a relatively larger punishment than the rich.

Deters the rich
- Another purpose of fines is to provide a deterrent. If fines are applied at a flat rate regardless of income, they must be low enough not to be unaffordable for those who do not earn much money. Consequently, they are set so low that they fail to have a deterrent effect on the richest in society, who are easily able to afford to break the law. A variable rate would avoid this problem.

SECTION TWO — 2008

Possible arguments against
Flat rate is more just
> A fine ought to be proportionate to the severity of the crime committed, not the income of the offender. It is fundamental that the justice system should treat all offenders equally; if two people commit the same crime in the same circumstances but one is richer than the other then they have caused the same amount of harm so should pay the same price for that harm. Having a richer person pay more implies that crimes by the rich are necessarily more harmful to society regardless of what the crime actually is.

Conclusion
> Summarize points made on both sides of the argument concisely and draw ends together.
> One option is to have an open ended conclusion – if one values justice then parking fines should not be based on income, however if one values the effectiveness of the law, then parking dines should be based on income.

"The cause of gender inequality is in the hands of men, but the solution is in the hands of women." Do you agree?

Although this question, at first, appears complex, the question is made easier to answer simply because the statement involved is extreme. As a result, and as we shall see below, it is very easy to take the statement at the heart of this question apart and analyse each section individually. As well as helping to create a more clear structure to your essay which is often difficult with these philosophical questions, it also makes it far easier to answer the question and make a convincing argument. However, in dissecting this quite extreme statement, it is very important that the conclusion you reach is not equally extreme in nature. It is tempting to simply refute the statement, as such a refutation is easy to make. What makes a better response is the formulation of a counter statement which you can argue better sums up the reality of the situation.

Introduction
> One possible approach is to contextualise the argument by firstly defining what is meant by gender inequality.
> One meaning could be a financial one – women are paid less than men for the same job. Another could be that it is an issue of power – there are fewer women in positions of power than men.
> If at all possible, introduce a figure from the news – for example only 32% of MPs are female, or for every £1 a man earns a woman earns 80p.

Argument

- In this case it is very importantly to look very closely at the statement. It is clear that there are two parts:
  1. The cause of gender inequality is in the hands of men
  2. The solution is in the hands of women

- A good approach would be to split the statement up into these parts and to then assess each part individually. (It is perfectly acceptable to agree with one and disagree with the other)

- A possible argument for 1) is that men occupy the majority of positions of power, and for 2) that it is only through campaigning for themselves that women will be able to overcome gender equality.

- Another approach could be to reverse and then critically assess the statements – i.e.:
  1. The cause of gender equality is in the hands of women
  2. The solution of gender equality is in the hands of men

- Possible arguments in favour of these statements could be that 3) gender equality still exists because women have not campaigned hard enough or 4) men dominate the power structure of society, therefore only they have the power to change it. (NB. these are both counter arguments to points 1) and 2).

- It is quite obvious that the way the question is worded leads to arguments which focus solely on one gender or the other either causing or giving the solution to gender inequality. How likely is it that this is the case? Another option could be that it is either the fault of neither gender, it is simply the result of biology, or that it is the result of both genders.

Conclusion

- Summarize the main points made on each side of the argument in the essay.
- Link back to the original quote – play with it if you need to.
- You may wish to come to a decision either way, or it is equally fine to sit somewhere in the middle, so long as this is fairly justified

END OF PAPER

2009

Section 1

Question 1: D
The argument discusses various reasons why people should be allowed to work beyond the current compulsory retirement age, and then strikes down a possible reason why the compulsory retirement age should continue to be enforced (the notion of struggling to get rid of inefficient elder employees). If we accept all these given reasons as true, it gives us good cause to accept the statement in D), thus answer D) correctly identifies the main conclusion.
Answers B) and E) are both reasons given in the passage to support this conclusion.

Answer A), meanwhile cannot be reliably concluded from the passage as the passage makes no reference to the fairness of the current compulsory retirement age. Answer C) is also irrelevant as the argument suggests the retirement age should be *abolished*, not raised.

Question 2: C
If 5cm is folded up on each side, the box is 40cm by 40cm on the bottom (50cm minus two lots of 5cm). It is 5cm high. Hence the volume is 5 x 40 x 40 cm, which is 8000cm^3. Hence the answer is C.

Question 3: D
The argument describes some biological tendencies of boys and girls, and how these may be causing the differences in performance between boys and girls in school. It then goes on to conclude that this means failing teachers cannot be responsible.

D) correctly points out that just because biological tendencies may be causative, does not mean they are entirely responsible for the observed phenomenon. Thus D) correctly identifies a flaw in the argument's reasoning.

A), B) and E) are irrelevant to how the argument's reasons lead on to its conclusions, so are not flaws. Meanwhile, Answer C) is not relevant because the argument is discussing how the inferior verbal and emotional skills of boys are causative, and makes no reference to the effect of their greater skills in calculating and planning.

SECTION ONE 2009

Question 4: B
The argument discusses how shortening the length of degrees would produce lower quality graduates, and then states that any move which results in lower quality graduates should be resisted by Universities. From this, it readily follows on that Universities should thus oppose the shortening of degree lengths from 3 years to 2 years, so B) is the correct answer.
Answers A) and E) are completely irrelevant to the argument's reasoning, so cannot be concluded. Answer C) is also irrelevant as the argument describes how universities should *resist* the move, making no reference to whether they are proposing it.
Answer D) is irrelevant because the argument refers to what *Universities* should do, not students.

Question 5: E
The passage describes how there are 2 options for treating depression: drugs, which alter the chemical balance of the brain, and psychotherapy. It then concludes that if we do not wish to alter the brain's chemical balance, we should choose psychotherapy. At no point is it stated that psychotherapy does not alter the chemical balance of the brain, and we can see that if this is not true, then this conclusion is no longer valid. Thus, E) is the answer.
A), B) and D) are all completely irrelevant, as the argument makes no reference to the causes of depression or which treatment is more effective.
C) is also irrelevant, since the argument refers to *if* we do not wish to alter the chemical balance, and it is stated that drug treatments do this, so the situation in C) is not relevant.

Question 6: C
Last time Jenny put petrol in the car, it had 5 litres in and she bought $6 worth of fuel at a cost of 60c per litre, which is 10 litres. Hence after filling it up, it had 15 litres in. She then drove 50km. Her car travels 100km on 8 litres of petrol, hence travelling 50km will use 4 litres of petrol. Hence the next time she goes to fill it up, it will have 11 litres in (15 minus 4). Since the tank holds 50 litres, Jenny will put 39 litres of fuel in (50 minus 11) if she fills it right up. At 50c per litre, 39 litres of fuel will cost $19.50. Hence the answer is C.

Question 7: E
To answer this question, halve the amounts in the "Period 1" column and compare these new amounts to the amounts in the "Period 2" and "Period 3" columns. We are looking for a row where half the amount in the period 1 column is less than the amount in both the "Period 2" and "Period 3" column. The only row where this is the case is the row for Belgium. Hence the answer is E.

SECTION ONE 2009

Question 8: D
Imagine the 3 sections of the shape from left to right in the original diagram are named x, y and z.
Shape A can be made by rotating sections x and z inwards in to meet in the middle. Shape B can be made by rotating section x in to meet the bottom of section y and then rotating the whole shape 180 degrees. Shape C can be made by rotating sections x and z round to point upwards.
Shape E can be made by rotating section z in to meet the bottom of section y and then rotating section x in to meet the side of section z.
Shape D cannot be made because if either section x or section z is rotated in to make the hexagonal shape, the other cannot then be in the position on the top due to where it is attached. Therefore the only shape that cannot be made is shape D. Therefore the answer is D.

Question 9: E
The passage discusses how Zoos do not showcase animals exhibiting normal behaviour, and then concludes that it would be preferable to abandon zoos, and instead use the money to protect natural habitats. If the statement in E) is true, then it follows on that the money spent on Zoos already boosts the populations of animals in natural habitats, through re-introduction of animals bred in captivity. This weakens how the argument's reasons lead on to its conclusion, as there now seems little point in abandoning zoos in order to boost/aid natural habitats. Thus E) is the answer.
D) is irrelevant, whilst A) actually strengthens the argument by reinforcing the notion that animals in cramped conditions become neurotic.

Question 10: C
The passage discusses how tests have shown that after intense use of memory, the brain sections responsible for memory are observed to increase in size, and thus the brain increases in size and power when used for a certain purpose. It then concludes that in order that taking part in many IQ tests will boost an individual's IQ. However, this is not a reliable conclusion because nothing in argument states that what applies to memory will also apply to general brain function. Thus, C) correctly identifies a flaw in the argument.
D) is irrelevant to the argument's conclusion. We do not need to have a certain number to accept the premise that many IQ tests will *increase* the brain's power, because no specific quantity of increase is given. E) is a direct refutation of the argument's conclusion and not a flaw. B) is irrelevant to the argument's conclusion, so it is not a flaw.

Answer A) is not correct because the argument has referred to multiple observations, with the London taxi drivers simply given as an *example*. Thus, we do not need them to be representative for the argument's conclusion to be valid, and A) is thus not a flaw.

Question 11: C
The passage discusses how many concerns about expert opinion are justified. Answers D) and E) are simply reasons/facts stated in the passage to support this conclusion, and are not conclusions in themselves. Answer A) is a statement of this conclusion.

The passage then describes how in many cases expert *evidence* is reliable, and not subjective to opinion, and how it would not be desirable for this to be discounted because of justifiable concerns over expert opinions. Answer B) is a statement of this conclusion.

Both of these conclusions go on to support the statement given in C), that we should therefore seek to distinguish expert opinion and expert evidence. Thus, A) and B) are *intermediate* conclusions, whilst C) is the *main* conclusion of this passage. Thus, C) is the answer.

Question 12: B
The round trip takes 4 ½ hours and Ionnais can walk twice as fast on the way back as on the way there. Therefore the trip back takes half as long as the trip there. If we denote the time taken to get BACK FROM the post office as t, then the time taken to get there is 2t. Hence the total time there and back equals $t + 2t = 3t$. Hence $3t = 4 ½$ hours, so $t = 1 ½$ hours. Hence we know that it takes Ionnais 1 ½ hours to do the journey at 4km an hour. Walking at 4km an hour for 1 ½ hours would result in a distance of 6km being travelled. Hence Ionnais lives 6km from the post office. Hence the answer is B.

Question 13: B

We can work out how much 100 minutes of call time will cost on each of the 15 packages given as follows:

	High User	Medium User	Low User
Activering	50 + (100 x 0.05) = £55*	20 + (100 x 0.30) = £50	100 x 0.60 = £60
Bellaphone	55 + (100 x 0.05) = £60	20 + (100 x 0.25) = £45*	100 x 0.60 = £60
Canconnect	60 + (100 x 0.05) = £65	20 + (100 x 0.35) = £55	100 x 0.55 = £55*
Dialfast	65 + (100 x 0.05) = £70	10 + (100 x 0.45) = £55	5 + (100 x 0.45) = £50*
Engaged	70 + (100 x 0.05) = £75	10 + (100 x 0.40) = £50*	100 x 0.70 = £70

The lowest cost package is the Bellaphone medium user package, hence the answer is B.

Note: Some of the options can be ruled out without calculation to save time.
For example, the high user packages for all providers apart from Activering are clearly more expensive than the Activering high user package given that the fixed charge is higher and the call charge is the same.

Likewise, the Engaged and Bellaphone medium user tariffs are the only medium user tariffs that need to be calculated because others have the same fixed rental but have a higher call charge.

The low user tariffs for Canconnect and Dialfast are the only low user tariffs that need to be calculated for the same reason. The packages that need to be calculated are asterisked in the table.

Question 14: E

We can write two simultaneous equations to work out how much food is needed for each small chicken and each large chicken. Let the amount of feed needed for a small chicken per day be S & the amount needed for a large chicken per day be L.
1) $100 = L + 2S$
2) $175 = 2L + 3S$

Equation 1 can be rewritten as $2L=200-4S$, and equation 2 as $2L=175-3S$. Therefore $200-4S=175-3S$.

We can rearrange this to $200-175=4S-3S$, hence $S=25$. Substituting this back into equation 1 elicits $100=L+50$, so $L=50$.

So a small chicken needs 25 pellets per day and a large chicken needs 50 pellets per day. Now let's look at the statements:

A is true because a small chicken needs 25 pellets per day and so 30 will last for longer than a day.

B is true because a large chicken needs only 50 pellets per day so 70 will be enough for one day

C is true because a large chicken needs 50 pellets per day and a small chicken needs 25, and 50 is twice 25

D is true because two large chickens would need 100 pellets (50 x 2) and four small chickens would need 100 pellets (25 x 4), making a total of 200 pellets.

E is false because three large chickens would need 150 pellets (50 x 3) and ten small chickens would need 250 pellets (25 x 10), making a total of 400 pellets.

Hence E is the only untrue statement.

Question 15: A

The passages discusses in depth how different types of dishonest acts are committed by different children for different reasons. From this, we can readily conclude that motives behind one child committing a certain dishonest act will not necessarily be there for a different dishonest act. Thus, we can reliably refute the assumption that a child dishonest in one situation may be dishonest in another, and thus A) can be reliably concluded.

Answers D) and E) are incorrect as the argument makes no reference to an overall reason underlying all dishonest acts. In fact, the argument refutes the existence of such a reason.

C) is also incorrect as the argument makes no claim as to what should happen in response to dishonest acts. The argument also makes no claim that *all* reasons for cheating are complex, so B) is incorrect.

SECTION ONE 2009

Question 16: E
The passage gives many reasons why the hard shoulders should not be utilised to relieve congestion, but then states that the main reason is to maintain Britain's good record of safety on motorways. At no point is it stated that this record is somewhat attributable to the hard shoulders, and if this is not the case this conclusion is no longer valid. Thus, E) correctly identifies an assumption in the argument.
All the other statements are irrelevant, because they have no effect on this conclusion that the hard shoulder should be kept open to maintain the safety record of the motorways. We can readily see that none of the other answers are required to be true for this conclusion to be valid. Thus, none of the other answers are assumptions.

Question 17: C
The passage discusses how speed-reading works, and states that this is of limited use for students as it does not allow a deep understanding of the text to be achieved. We can see that the statement in C) contradicts the suggestion that reading at speed does not lead to a deep understanding, and in fact suggests the opposite. Thus, C) weakens the argument's conclusion.
Answers A) and D) both reinforce the suggestion that speed-reading does not lead to a sufficient understanding for students, and thus these would actually strengthen the argument's conclusion.
Answers B) and E) are irrelevant, as they do not refer to how speed-reading affects a person's understanding of a given text.

Question 18: B
The boy spends at least half his money, so 50c, on gobstoppers. He then spends at least a quarter, so at least 25c, on fruit chews. Fruit chews are 3c each so the minimum he spends on fruit chews is 27c. This leaves 23c for him to spend on other things. We know that he spends at least a tenth of his money, so 10c, on bubble gum. Therefore he spends between 10c and 23c inclusive on bubble gum. At 2c each, this is from 5 to 11 pieces. Hence the answer is B.

Question 19: D
We can go through the calendar and see how many consecutive days in any of the countries are public holidays by observation. From January to March, the longest period of consecutive bank holidays is 5 days. This stretch of holidays occurs from May the 9th (which is a Holiday in Denmark) to May 13th (A holiday in several countries). We see that each day between these 2 dates is a public holiday in at least 1 country. Hence, the Answer is D)

Question 20: A

We can see that Pulley A is half the size of Pulley B. Thus, for every 2 complete rotations of A, B will rotate once if rotating at the same speed (As determined by the Belt connecting Pulley A and the outer pulley of Pulley B)

This means the inside pulley of Pulley B will also rotate once for every 2 complete rotations by Pulley A (since both parts of Pulley B will rotate at the same speed).

We can see that Pulley C is 4 times bigger than the inner Pulley of Pulley B. Thus, for every 4 rotations by Pulley B, Pulley C will rotate once if rotating at the same speed.

Thus, in order to get the number of complete rotations by Pulley C, we divide the number of rotations of Pulley A by 2, then divide this number by 4.

Thus, for every 8 rotations of Pulley A, Pulley C will rotate once. Hence the answer is A).

Question 21: D

The passage discusses how the widespread prevalence of the English language is given as a reason for the poor performance at learning languages, since they have little need to speak other languages abroad. It then refutes this reason, saying that even amongst children who have never been abroad the English are poor at learning languages. It then concludes that this must mean the reason is actually that English schools are poor at teaching languages.

Answer D) correctly points out that there may be a whole host of other reasons behind this phenomenon, so the argument's conclusion is incorrect. Thus D) is the answer.

Answer B) actually strengthens the argument, by suggesting other languages are also widespread, thus reinforcing the notion that the widespread nature of English is not a reason for the poor performance of the English in learning languages. A) is irrelevant, as the reasons behind why English schools may be deficient at teaching languages (such as poor funding) do not affect the fact that they are deficient. E) is also irrelevant as the fact the English are poor at learning languages is a stated fact in the argument, which we thus accept as true for the purposes of a critical thinking assessment. Hence, it does not require proof in this context.

Answer C) is incorrect because the sample is given to refute the notion that lack of requirement to speak foreign languages abroad are the reason behind the poor performance of the English. Thus, the sample does not need to be representative of the population in order to be valid for its purpose.

SECTION ONE   2009

Question 22: D
The passage's reasoning can be summarised as "If A happens, B will happen. If C happens, D will happen. We need D more than we need B, so C will happen". We can see that answer D) follows this same reasoning, with "B" being a pleasant house, and "D" being a pleasant me.
Answer A) differs in that it is stated we do not care about losing weight, not that one option is preferable to the other, thus we will follow the other option.
Answer B) differs in that learning to drive is stated as being needed, but claimed that this can be learned later. This is different from simply claiming it is less preferable/needed than the other option, as in the question.
Answers C) and E) both refer to negative consequences of *not* following one of the options, and thus are different from the question, which simply refers to a choice between 2 positive consequences.

Question 23: A
The passage discusses how users of mobile phones on trains and in restaurants would object if other passengers/customers were to cause noise and disruption to them, and therefore they should not cause disruption to others by using their phones. The principle here is that you should not do something to others that you would object being done to you. Answer A) follows this principle, that parents should not do to their children things that they would object having done to them.
Answers B), C) and E) are completely irrelevant to this principle, whilst Answer D) actually describes an opposite principle (namely that we should accept these annoyances for the sake of other's convenience).

Question 24: B
If Roger and the goalkeeper play the whole match, then there are 9 players on the pitch who are ones who are being "rotated". If there are 16 players in total, without Roger and the goalkeeper this is 14 players. Hence each player will play 9/14 of the whole match, which is 70 minutes. 1/14 of a 70 minute match is 5 minutes as 70 divided by 14 is 5. Hence 9/14 of the match is 45 minutes. Hence the answer is B.

Question 25: C
The question is how far behind June's time zone April lives, because we want to know how much later April thinks 23:30 on Wednesday is than June. April lives 5 hours behind London time and June lives 9 hours ahead. Hence the total time differences is 14 hours. Hence the answer is C.

Question 26: D

The distances from Nickel to the other towns are: 28km, 40km, 63km, 68km, 72km, 102km and 119km.

If the distance from Nickel to another town is 28km, it must be at the end of one of the 28km long roads as no other combination of roads adds up to 28km. Hence Nickel must be P, Q, W, V, S or T.

If the distance from Nickel to another town is 40km, it must also be at the end of one of the lengths of road where a 28km road is connected to a 12km road, as no other combination of roads adds up to 40km. Hence Nickel must be W, S, V or T.

If the distance from Nickel to another town is 63km, it must be also at the end either of the 63km road or at the end of the combination of connecting roads of length 23km, 12km and 28km. Hence Nickel must be T or W.

The distance from T to W is 68km so using this criteria the answer can still be either T or W. However there is no combination of roads from W that adds up to 72km. However, T to R is 72km. Hence the only town that can have this combination of distances to the other towns is T. Hence Nickel is town T. Hence the answer is D.

Question 27: E

The passage discusses how over-qualification is often cited as a reason for rejection, and then argues that there are no circumstances in which someone is actually over-qualified, and thus it should not be cited as a reason for rejection on a job application.

Answers A) and E) are both valid conclusions from this passage, which readily follow on from the reasons given. However, we can see that the statement in A), if true, goes on to support the statement in E), which is also a valid conclusion. Thus, A) is an intermediate conclusion, whilst E) is the main conclusion of this passage.

B) and C) are both restatements of reasons given in the passage, and are not conclusions. They do both go on to support the conclusions given in A) and E).

We can see that D) is required for the conclusion in E) to be valid. If D) is not true, then the fact that nobody can be overqualified for a job does not necessarily mean that employers should not list this as a reason for rejection. However, D) has not been stated. Thus, we can actually see that D) is an assumption in the passage, not a conclusion.

SECTION ONE — 2009

Question 28: D

The passage discusses how the generation born in the 1940s/1950s will have a good chance of living a long lifespan, due to a number of healthy habits they indulge in. It then discusses how the generation currently in their 20s/30s indulge in a number of unhealthy habits, and are thus likely to suffer from health problems and die sooner. We can see from this summary that we have only been given information on 2 generations, and thus we cannot conclude what trends are present among life length for different generations, and how these are likely to change on the whole. Thus, the statement in C) cannot be concluded, as it refers to a trend of "ever-increasing life expectancy", which we cannot conclude exists based on the information in the passage.

B) and E) are completely irrelevant as the passage makes no reference to the causes of such healthy/unhealthy habits, or the effects on the health service that will follow. Thus, B) and E) cannot be concluded.

However, we *can* conclude that it would be wise for the "big mac generation" to change their habits, as the passage has discussed how these will lead to decreased life span and more health problems. Thus, Answer D) can be reliably concluded.

Answer A), meanwhile, is actually an assumption in the passage. We can see that if this is not true, then the argument's conclusions are no longer valid. Thus, A) is an assumption and not a conclusion.

Question 29: B

The passage discusses how the law should be changed to allow people to sell their kidneys, because this would result in a greater number of successful kidney transplants. We can immediately see here that this relies on an assumption that some people would be willing to sell their Kidneys, and if this is not true, this conclusion is no longer valid. Thus we can readily see that B) is an assumption. Thus, the answer is B).

The passage then goes on to discuss why we should not be dissuaded by fears of rich patients exploiting poor donors, as this is no different from working in a dangerous occupation. Since the primary reason for changing the law has already been given, this additional discussion is no longer *required* to be true for the argument's conclusion to be valid. Thus, C) is not an assumption, as it is not required to be true for the conclusion to be valid.

D) and E) do not affect the conclusion. Whether foreign transplants are successful or not does not affect whether more donors in the UK would result in more successful transplants. Equally, the argument is referring to there being *more* successful transplants, so the notion of whether an adequate supply would be attained is irrelevant.

Answer A) is a completely irrelevant statement.

SECTION ONE 2009

Question 30: E
If the month of January contains five Fridays, then it must either start on a Wednesday, Thursday or Friday and hence finish on a Friday, Saturday or Sunday. You can see this by writing out the possible combinations of 31 consecutive days starting on each day of the week and counting the Fridays as follows:
MTWTFSSMTWTFSSMTWTFSSMTWTFSSMTW
TWTFSSMTWTFSSMTWTFSSMTWTFSSMTWT
WTFSSMTWTFSSMTWTFSSMTWTFSSMTWTF
TFSSMTWTFSSMTWTFSSMTWTFSSMTWTFS
FSSMTWTFSSMTWTFSSMTWTFSSMTWTFSS
SSMTWTFSSMTWTFSSMTWTFSSMTWTFSSM
SMTWTFSSMTWTFSSMTWTFSSMTWTFSSMT
Hence A is definitely true, as Friday must either be the 1st, 2nd (if the month starts on Thursday) or 3rd (if the month starts on Wednesday). B also could be true because if the month starts on a Wednesday, there will be 5 Fridays (3rd, 10th, 17th, 24th and 31st). C is also definitely true because no matter which of the starting days of Wednesday, Thursday or Friday we pick there will only be 4 Tuesdays. D also could be true as if the month starts on a Friday, the 31st will be a Sunday and hence there will be 5. However, E cannot be true because if the 31st is a Monday, there are only 4 Fridays (7th, 14th, 21st, 28th). Hence the answer is E

Question 31: C
For the 45 weeks the restaurant is open, there are total overhead costs of £45000 (£36000 rent + £9000 other expenses). Hence the overheads cost £1000 a week. The total wages per week are £1200. Hence the total costs paid by the restaurant for all non-food expenses are £2200.
During each week the restaurant is open, it is open for 10 sessions at which on average 20 people eat, so 200 people eat there each week. The food for their meals costs another £1000. Hence the total expenses to the restaurant are £3200.
Splitting £3200 between 200 people is £16. Hence the owner must charge £16 for each meal to cover expenses. Hence the answer is C.

Question 32: C
➢ In map A, 2 stations have 2 connections and 3 stations have 3 connections.
➢ In map B, 2 stations have 2 connections and 3 stations have 3 connections.
➢ In map C, 1 station has 2 connections, 3 stations have 3 connections, and 1 station has 4 connections
➢ In map D, 2 stations have 2 connections and 3 stations have 3 connections.
➢ In map E, 2 stations have 2 connections and 3 stations have 3 connections.
Hence the one map that must be wrong is map C because it shows different connections to the others. Hence the answer is C.

SECTION ONE   **2009**

Question 33: A

The passage discusses the timeframe in which batsmen observe a bowled ball and respond with a shot. It then discusses how science claims this is an impossible timeframe in which to react like this, concluding that this means science must be wrong, and failing to explain a phenomenon due to inherent limitations.

However, if we accept the statement in A) as true, then suddenly the claims of scientists seem much more reasonable. It may well be that there is not enough time to consciously react, and that instead a reflex response has to occur, which is much faster than a conscious response. Thus, Answer A) weakens how the reasons in the passage support its conclusion.

Answers C) and D) both strengthen the argument, as C) reinforces the notion of people reacting in timeframes which science claims are impossible, whilst D) reinforces the notion that science may have gaps and limitations which reduce its ability to explain phenomena.

E) is irrelevant as the claim is that such a reaction is *impossible*. The fact that it only happens *sometimes* does not affect the erroneous nature of this claim.

B) is completely irrelevant. It does not matter how the information was found out, this does not affect how correct the claims of scientists are about it.

Question 34: D

The passage argues that education standards are failing, and that government claims of improving standards are driven by tests simply becoming easier. The main evidence to support the notion that education standards are failing is that in recent tests almost half of students attained below average results.

However, D) correctly identifies that this is flawed evidence. If we calculate the average of the scores in a given test, we would expect it to be somewhere around the middle, with roughly half of people performing above-average, and roughly half performing below-average. Thus, this figure does not say anything about the quality of the test of the levels of attainment that students achieved. It merely confirms what we would expect due to how statistics work.

A) simply states the other side of the statistic given in the test, and thus does not affect the argument's reasoning.

B) and C) are irrelevant to the notion of whether standards are falling, and whether we need to return to teaching methods of the 1950s, so are not flaws.

E), meanwhile, correctly points out that there could be other explanations for falling standards, but is not a valid flaw. Just as we cannot conclude falling standards are due to failing teaching methods, neither can we conclude that they are due to changed social attitudes. If E) stated that the falling standards *could* be due to differing social attitudes, then this would be a valid flaw. However, since E) presents a fact that cannot be *certainly* concluded, it is not a valid flaw.

Question 35: B

The passage discusses how we may *admire* certain acts, but cannot refute that they are morally wrong. It argues that many acts (such as those stated) would be described as morally wrong by anyone who knew what this phrase meant, and thus some things can be shown to be morally wrong as much as the snow can be shown to be white.

From this, we can reliably conclude that the statements in B), C) and E) are true. Thus, these are all valid conclusions from this passage. However, we see that Answers C) and E) both go on to support the statement in Answer B). Thus, C) and E) are *intermediate conclusions* from the passage, whilst B) is the main conclusion.

Answers A) and D) are *not* valid conclusions from the passage. The passage discusses how certain acts are *definitely morally* wrong, but it is *possible* to admire them and think that they are right. Thus, we cannot conclude that such acts are *wrong*, only that they are *morally* wrong. Equally, there is nothing in the argument that means we can definitely state that they are *right*, we can only conclude it is *possible* to take this view. Thus, neither A) or D) can be reliably concluded from the passage.

Question 36: B

If 50 calories would be used from rowing 500m, then rowing 400m would use 40 calories (50 x 4/5)
If 55 calories would be used from rowing 500m, then rowing 600m would use 66 calories (55 x 6/5)
For the third session, 60 calories were used for the first 500m. For the remaining 300m: if 40 calories would be used from rowing 500m, then rowing 300m would use 24 calories (40 x 3/5).
Hence the total calories used is 40 + 66 + 60 + 24 = 190.
Hence the answer is B.

Question 37: D

The only factor relevant to whether the tree will obscure the block of flats is its height. The question also refers to *when the tree is fully grown*, thus growing time is irrelevant. Thus, we can safely ignore the columns detailing the features, growth speed and width, and just focus on the height of the trees. This simplifies the question considerably.

SECTION ONE 2009

We can see that a bush 1.6m in height obscures a building 7m tall.

The new building to be obscured is 20m, which is roughly three times the height of the Bungalow. We know that the distance away from the observation point, and the observation point itself have remained the same. Thus, we know the new tree must be roughly three times the height of the bush to obscure the block of flats.

Three times 1.6 is 4.8. Thus, we look for a tree which is at least 4.8m high. We also know that Mrs Brown wishes the tree to be as small as possible whilst obscuring the flats, so we are looking for the smallest tree which is at least 4.8m high when fully grown. This is the Hornbeam. Thus, the Hornbeam would be the most suitable tree.

Question 38: D

We cannot infer anything about Northern Irish house prices from this graph unless we are given information about UK house prices as whether they are true or false depends on Northern Irish house prices not on the ratio of Northern Irish to UK house prices, so we cannot say for sure that A, B or E are true. We also cannot say that C is true because it is impossible to infer anything about UK house prices from this graph without knowing what happened to Northern Irish house prices. We can however say that D is definitely true because in some years, the gradient of the graph was positive (the line goes upwards) which means that in that year Northern Irish house prices rose as a percentage of the UK average in that year. Hence the answer is D.

Question 39: A

The passage states that policies to prevent drink driving often rely on severe penalties as a *deterrent*. It then claims that such policies are only a deterrent *if* people perceive a moderate or high chance of being caught for these offences, and that people *do not* perceive the chances of being caught are moderate or high. Thus, the *if* has not been satisfied, and we can readily conclude that severe penalties do not constitute an effective deterrent. Thus A) is the answer.

B) is not correct because the passage only claims that governments have *tended* to rely on severe penalties. Thus we cannot conclude that they have **not** tackled the problem of drink-driving.

C) and E) are incorrect as the passage makes no reference to what the actual chances of being caught are, or whether other policies would be more effective. Stating that one policy is not effective does *not* necessarily mean that another will be effective.

D) is incorrect because the passage has discussed how severe penalties are not having an effect due to low perceptions of the chances of being caught. Thus, it follows that even more severe penalties would also be ineffective.

SECTION ONE   2009

Question 40: C

The key to this answer lies in the passage's claim that "the **only fair** solution" is to take a levy from the pay of top athletes, to foster the development of up-and-coming athletes. At no point is it stated that the general taxpayer subsidising development of young athletes is **not** fair, and if this is not true, then suddenly we are presented with another **fair** option, and thus the argument is no longer valid. Thus, the argument relies on the assumption stated in C).

Answers A) and D) are incorrect because the passage simply claims that some athletes will fail due to lack of money if we do not take a levy from top athletes. How many will fail is irrelevant to the validity of this statement, so A) is incorrect, whilst D) is incorrect as the passage has *stated* that some will fail so we must accept this as true, and thus possible solutions for up-and-coming athletes are not relevant.

Statements B) and E) are irrelevant as the passage does not refer the notion of athletes becoming winners (it simply states that we should help not to fail due to money issues) or whether top athletes deserve their pay (it merely states they expect it as a reward). Thus, neither of these answers affects the **argument's reasoning or its conclusion.**

Question 41: D

The passage discusses how children whose parents act aggressively towards them often grow up to be violent, and concludes that if this were stopped we could eliminate a lot of violence from society.

The passage refers to what would happen *if* we could stop aggression towards children, so any comments on how difficult/impossible this is are irrelevant. Thus A) is incorrect.

C) and E) are irrelevant. What the distinction is between reasonable discipline and aggression, and the causes of aggression are irrelevant to the effects that aggression produces, so neither of these answers affect the argument.

B) does not weaken the argument because it makes no reference to the effects of aggression towards children. The fact that children of aggressive parents are not themselves aggressive does not affect how aggression in the first instance produces violence in people. Thus, B) does not affect the argument's conclusion.

D) does weaken the arguments as it states that other factors are *the major* causes of aggressive behaviour in both adults and children. If this is true, then it suggests prevent aggression from parents will not affect the major reasons behind aggression/violent tendencies in children, thus weakening the conclusion that this would eliminate *a significant proportion* of violence.

Question 42: E
There are 16 multiples of 3 between 1 and 49 inclusive (3, 6, 9, 12, 15, 18, 21, 24, 27, 30, 33, 36, 39, 42, 45, 48).
There are 9 further numbers which contain a 3 (13, 23, 31, 32, 34, 35, 37, 38, 43)
There are 3 further numbers which contain a 6 (16, 26, 46)
There are 3 further numbers which contain a 9 (19, 29, 49)
Hence the total amount of numbers that fit one or more of the criteria is $16 + 9 + 3 + 3 = 31$

Question 43: B
Oliver will be leaving on the 10th anniversary of beginning to work for the company. Thus, this will be at the end of his 10th year of working for the company. The allocation for this year is 25 days holiday. Thus, this will be the amount granted when Oliver hands in his notice, in order to stave off bad publicity.
We can see that in his 9th Year, Oliver took 20 days holiday time, which is 5 days less than the entitlement for the year, 25 days. Thus, 5 days will be carried forward to the 10th year.
Thus, a grand total of 30 days holiday will be allowed. Subtracting the 5 days of holiday already taken, we can see that this will result in 25 days *more* being allowed before he leaves.

Question 44: C
Over a wide area, the pattern can be considered a repeat of the section below.
This pattern contains 1 hexagon and 12 triangles.
Hence the answer is C.

Question 45: A
The passage discusses how agriculture and conservation have conflicting interests in wetland areas, with agriculture wanting to drain the wetland, and conservation wanting to not drain the wetland areas. The passage then acknowledges the conflict here, and says that because of this, an independent body should regulate wetland areas to achieve a balance.
We can readily see that the different intentions (draining and not draining) are mutually exclusive, so are irreconcilable. Thus, C) is a valid conclusion from the passage. B) and A) are also valid conclusions from the passage, and we can see that B) and C) both go on to support the statement in A). Thus, A) is the main conclusion, and B) and C) are *intermediate conclusions* in this passage.

SECTION ONE 2009

D) is actually an assumption in the passage. At no point has it been stated that wetland *needs* to cater for both interests, and if this is not true we can no longer conclude that a balance must be struck – we could simply ignore one of the two opinions. Thus, D) is an assumption, and not a conclusion.

E) is an irrelevant statement, which is not a conclusion and is not *required* for the argument to be valid, so is not an assumption either.

Question 46: A

Time spent travelling can be viewed as the direct result of speed of travel and distance travelled. We can see here that if either of these two factors changes, it will directly affect time spent travelling. Thus the passage's reasoning describes a circumstance where one factor (Factor Z) is a direct result of 2 others, and that Factor Z is remaining constant. We are then told that if one of the factors contributing to factor Z is changed, the other factor must also change accordingly, such that Factor Z remains constant.

Only answer A) follows this pattern. Here, Factor Z is the mortgage as a proportion of annual income. This will obviously be directly affected by the size of the mortgage and the annual income in question. Since Mortgage as a proportion of Annual income is remaining around the same level, if annual incomes change, the size of the mortgage must change accordingly. Thus A) is the answer.

B) is not the same reasoning because there are many other factors in "total expenditure per household" than food. Thus, if food becomes more expensive, it *does not* mean that total expenditure will increase, so we cannot conclude that mortgage repayments will also increase.

C) is incorrect because inflation is not dependent on the rate of a mortgage repayment. Thus, we cannot include the fixed rate mortgage repayments in a calculation of how inflation will change.

D) is incorrect because there is a distinction between the number of homes present and the number of homes *owned*. Thus we cannot draw conclusions about things which are linked to home ownership from the building of new houses.

E) is incorrect because the number of old houses being renovated is not necessarily linked to the number of old houses on the market. More old houses could be put on the market without renovation, so E)'s conclusion is not valid.

SECTION ONE 2009

Question 47: A

The passage refers to a situation where there is a *small* risk of a negative consequence occurring. However, *nothing can be done* about the risk, and informing people would cause unnecessary distress.

Only Answer A) follows this principle, where the risk of damage is small and nothing can be done about it, and informing people of the possible risk would cause worry and distress.

B) is different because in this instance, there is already a confirmed negative consequence, not a small risk of one. This is different from the principle in the question. C) follows a similar pattern, the train *is* delayed, so the negative consequence has already been confirmed, which is different from there being a small risk of a negative consequence.

D) seems to be correct; as the child may or may not be distressed by finding they are adopted. However, this is different from the principle in the question. In D), the event of adoption has already occurred, whilst the risk is causing distress. This is different from *definitely* causing distress by informing about *the risk* of an event occurring.

E) is different from the question because something *can* be done about this negative event. The child could implement new working strategies etc, so this is not the same as a situation where nothing can be done about a possible negative consequence.

Question 48: C

We can picture that in order for a driver to return to X during the shortest possible journey, some of the towns must be in directly opposite directions from X. Thus, it would be quickest to return to X, between visiting these 2 towns.

We can also see that this can only be the shortest journey if the third town is located on the opposite side of X from one of these two towns, such that a straight line drawn from X would go through both these towns. (Otherwise it would be quicker to head from the town on the opposite side of X straight through to this third town). If this is the case, then the quickest route is to visit the town opposite X from these two, then return to X, then visit the second town, then the third town, *then visit the second town again* on the way back to X.

We can see that route C must have this setup of towns, but does not visit B again after visiting C. Thus, C) cannot possibly be the shortest route.

SECTION ONE — 2009

Question 49: B

First we must calculate how much rainwater actually falls on the roof. We are told that 160 litres falls on every meter squared of roof, and that there are 25m² of roof in total. Thus, a total of 4000 litres of water fall on the roof (160 times 25 is 4000). We are told that half of this is collected, so a grand total of 2000 litres of rainwater is collected. We are also told that there is never enough rainwater to overfill the butt, so we can assume that all of this 2000 litres is used to refill the pond.

Now we need to calculate how much water Joan uses. This is straightforward. She uses 100 litres a week, for 25 weeks, giving a grand total of 2500 litres of water used. Thus, Joan has collected 2000 litres and used 2500. Thus, there is a 500 litre deficit between the rainwater collected and the rainwater used.

However, we must also factor in the 200 litres of water that was present in the butt at the beginning. When this is included, we get a grand total of 2200 litres of water from the butt used to fill the Pond. 2500 litres was actually used, so Joan will need to use 300 litres of mains water. Hence, the answer is B).

Question 50: E

The view shown in E) is not a possible view because the Red and Yellow sides are in the wrong positions.

We can see from the view given, that when Blue and Yellow are folded round, they will be positioned such that if we are still looking at Red Face-on, the blue will be on the *left* behind the red, and the yellow will be on the *right* behind the red. When we rotate this round to get a top view, we see that if blue is on the bottom, then Yellow must be on the *right*, and red must be on the *left*. Thus, View E) is not a possible view. All other views are possible.

END OF SECTION

Section 2

Albert Einstein wrote that "The whole of science is nothing more than the refinement of everyday thinking." Do you agree?

The essay is easy to structure well; clearly define the rather loose terms 'science' and 'everyday thinking', then begin with the arguments for the case you want to make (that is, in favour of or against Einstein's sentiment). This should naturally be followed with counterarguments these points, and any additional arguments for or against (whichever you have not chosen to support), and then counter-counterarguments to these, and a reaffirmation of your position, which can be made in the conclusion.

Introduction:
Explore what Einstein meant by this quote:
- What constitutes 'everyday thinking'? This could be described as the way in which people navigate the world in their general lives – their reasons for believing what they believe and acting as they do.
- What constitutes 'science'? You may wish to define this in terms of an active investigation into truth, or an application of the scientific method which answers questions by making observations and drawing conclusions consistent with all that has been observed.
- 'Refinement' is also an interesting word – if you argue that there are differences between everyday thinking and scientific thinking, is 'refinement' the right word to describe it?
- Einstein is asserting that the way you think in a scientific investigation is an extension of the same logic you already use in your thinking – consider whether you think this is true, or if there are elements of a scientific investigation that are counter to our usual way of thinking.
- Clearly state whether you agree or disagree with Einstein and outline your reasons, as in the form "I believe X, because Y; although one could argue Z, this is not the case because W". Though of course this exact formula need not be used, you should make your position similarly clear.

Potential arguments for:
1. The logic used in the scientific method is much the same as the logic used in everyday life. For example, we generally do not accept that an assertion should be considered true unless there is some kind of evidence to support it.
2. *As a response to the second argument against, listed below*: the methods used in scientific research are not equivalent to the 'thinking' that Einstein meant. The 'thinking' constitutes the reasoning and conclusions we draw from evidence, not the evidence itself or our method of acquiring it.

SECTION TWO 2009

3. The answers we seek to answer in scientific investigations are largely the same questions we have in daily life – why does X behave as it does? Why does Y happens? How do I get Z to happen? People are naturally curious about the answers to these questions from childhood; science is the 'refinement' of this thinking in that it provides a structure and method to the investigation.

Potential arguments against:
1. *As a response to the first argument for*: though it is true that we generally need a reason to believe something, what constitutes an *adequate* reason according to the mainstream scientific method is, or can be, vastly different from what a layperson considers an adequate justification for some other belief.
   a. Counterargument: this is why Einstein has used the term 'refinement'; we accept in everyday life that people should have reasons for believing what they do, but in a scientific investigation we define the terms of what constitutes a good reason.
2. The methods used in scientific research are far removed from the way in which we draw conclusions in everyday life; the use of carefully controlled experiments carried out with the aim of producing evidence for or against a given thesis cannot be considered 'everyday thinking'.
3. Some branches of science require thinking that is highly counterintuitive. Physics concerning very small particles, for instance, is like nothing we encounter in everyday life.
   a. A counterargument to this is that this is not concerned with the thinking about it. While we typically wouldn't think of objects interacting in the way that very small particles do, this is not the 'thinking' that Einstein meant; he meant our thinking is naturally investigative and rational, and these two qualities still apply when we are launching scientific investigations into not very everyday things. We still investigate and use our observations to draw conclusions, even if these conclusions seem bizarre at first.

Conclusion:
➢ Restate your position and summarise your main arguments.
➢ Be sure to closely link your conclusion to the original quote, mentioning scientific methods and everyday thinking.

SECTION TWO — 2009

If 'Humanitarian Intervention' is acceptable, why shouldn't Europe invade the USA to stop it using the death penalty?

This question can be read as asking two things – if 'humanitarian intervention' is acceptable, and, if it is, whether it is acceptable in this situation. As such, you should be sure to engage with both questions, asserting your position clearly in the introduction and summarising it in the conclusion. This essay can focus in depth on either the ethical or political considerations posed by the question, or can deal with both approaches slightly more superficially.

Introduction:
- Clearly your opinion in the introduction, and outline your reasons for holding it, as well as the counterarguments you will present and your reasons for dismissing them.
- Define 'humanitarian intervention' – a state's use of "military force against another state when the chief publicly declared aim of that military action is ending human-rights violations being perpetrated by the state against which it is directed."
- Consider whether humanitarian intervention is *ever* acceptable. If you are going to argue that Europe shouldn't intervene in the USA, you should make it clear whether this is because humanitarian intervention isn't ever acceptable, or it isn't acceptable in this case.

Potential arguments for intervention:
1. The death penalty is immoral and a human rights violation. Arguments for this include:
   a. Every person has the fundamental right to life
   b. European countries have outlawed the death penalty and declared life to be a fundamental right, show they must be of the opinion that the death penalty is a breach of rights
   c. Death is not an acceptable punishment for crime because the justice system is fallible, and this sanction is irreversible.
2. It is our duty to prevent the loss of life and human rights violations where possible.
3. It is unfair to apply foreign policy inconsistently; that we do not intervene when we see injustice in the USA is a result of Britain's desire to maintain our 'special relationship' with the USA, rather than a sound ethical judgement.
4. Intervention in the USA would set an example by showing the world that human rights violations are not acceptable. Additionally, the USA is equally as invested as Europe in keeping our relationship as peaceful as possible, so it is more likely they would be prepared to use methods other than

indiscriminate violent military action in response. Our mutual need means intervention in the USA is more likely to be productive that intervention elsewhere.

Potential arguments against:
1. The political and military unrest that an attack from Europe on the USA would cause would have consequences far more devastating and lethal than the death penalty is currently having.
2. Military intervention is not an acceptable way in which to address grievances with another country's conduct, or should be used only as a last resort in order to minimise violent conflict. Instead, European governments could petition the USA to reconsider its policy, or use our mutual dependence as leverage by threatening economic sanctions if the death penalty is not revoked.
3. The death penalty is not a human rights violation and is acceptable. Arguments for this include:
   - Some crimes are so horrific and damaging to society that they cannot be forgiven, nor can there be any attempt at rehabilitation.
   - The death penalty acts as a deterrent and prevents reoffending, and so decreases violent crime overall.
4. The USA is a democracy, meaning the laws that uphold the death penalty were instated by politicians given power by the electorate; as foreigners, for us to launch an intervention is to undermine democracy and their right to self-determination as a nation.
   a. A counterargument to this is that democracies are vulnerable to a tyranny of the majority. If there are not adequate safeguards to protect a vulnerable minority, it is our moral duty to aid them.
5. The use of the death penalty in the USA is relatively rare and used only in response to violent crimes; we should focus our efforts in countries where it is more common and used as punishment for 'crimes' such as civil disobedience or homosexuality.

Conclusion:
- Restate your position and your main arguments for it, as well as why you've not agreed with your counterarguments.
- Link it back to the question by specifying whether humanitarian intervention is in fact acceptable in any case, and then whether it is acceptable in this situation.

If you can give reasons for your actions, does that mean that your actions are rational?

The major argument in this essay is over the definition of 'rational' and whether it should include a criteria for a good or correct reason for acting; it is around this argument that it is most advisable you structure your essay.

Introduction:
- Give a broad outline of the definition of 'rationality' that you are going to be arguing for and give a broad outline of your reasons for doing so. Also introduce the counterarguments to your view, and your reasons for dismissing them.
- It is worth giving a brief explanation of how you will be using the word 'reason'. You will probably want to establish that the 'reason' must be true, it cannot simply follow the formula "I performed X action because Y", but rather the agent must truly have acted because of the fact of Y, or because they believed the fact of Y (and this distinction, as we will see, may be important later in the essay).

Potential arguments for:
- Rational action consists of actively pursuing a goal by acting in the way you think will enable you to achieve it. For example, if my goal is to quench my thirst, I drink water because I believe this will stop me being thirsty. Further, this follows even if my belief is false; for example, if my goal is to quench my thirst, and I brush my hair *because I believe this will stop me being thirsty*, I was rational even though I was wrong.
- Rationality means doing the most advisable thing given the resources you have – this includes information, mental faculties, and potential courses of action. If, from the resources available to you, you have drawn, say, the conclusion that brushing your hair will quench your thirst, then it makes sense to perform this action.
- The example used is deliberately absurd to demonstrate that the way of thinking, rather than how correct the belief is, is what is important to rationality. You may demonstrate that, in fact, this follows from how we view more typical cases of rationality in situations of false beliefs.

- For example: I want to catch the train at 4:30. I think it will take me an hour to get to the station, and I want to leave some time spare in case of a delay, so I leave at 3:00. A major accident happens while I am on my way to the station, delaying me by over an hour, and so I miss my train. In this situation, did I act rationally in leaving at 3:00? Most people would say yes, on the basis that although it later transpired that I had not left adequately early in order to catch my train, I chose the leaving time most recommended according to the information available to me – my beliefs regarding how long it would take me to get to the station and my beliefs regarding how much extra time I should leave in case of an emergency. Once we have established this as the definition of rationality, we can extend it to demonstrate that even in the more absurd scenarios, it follows that giving a reason for your action is all that constitutes rationality.

Potential arguments against:
The criteria for rationality is more stringent than in the argument given above – you must be able to justify that your *reason* was rational, not just that you *have* a reason.

> If you reason must be rational, there is the possibility of starting an infinitely regressing chain of explanations. When I say I perform X action to achieve Y goal, I must ask why I believe X will achieve Y. I give the answer Z. Why, then, do I believe Z means that X will help me achieve Y? The question here is whether these subsequent questions constitute an element of X action's rationality, or whether they are questions regarding a completely different rationalisation.

> If you take the former side – that they are necessary questions for establishing the rationality of X – then you may use this as an argument for answering 'no' to the essay question, but you are left with a further question you must answer: how far back in this chain do you have to go to establish rationality? Since this is a very obvious counterargument, if you make this argument you must be prepared to address it.

> You may give an answer to this counterargument along the lines of "you must rationalise your reason for X by one step, anything further than that is irrelevant", but this is weak to two further counterarguments:
  - The line you draw between what is relevant and what is not is arbitrary.
  - You may respond to this counterargument by reasserting that we are asking whether X action is rational, and giving Y reason. To ask whether Y reason is rational is a different question. This is why there are two

levels of justification – the reason for performing the action, and the reason the reason is a good one – and not any further levels.
- However, this argument is vulnerable to counterexamples such as: "I drink water to quench my thirst. I believe drinking water will quench my thirst because some dogs are brown." Since you are not required to justify the rationality of your reason for X, this would have to be considered valid. This is the case for any arbitrarily drawn line – you do not have to justify the rationality of your final reason, and so it can be utterly nonsensical and still be acceptable.

➢ A 'reason' alone is not enough to establish rationality; there are qualifications on what this reason must be in order to be sufficient. One condition is that you must be correct in thinking that Y action will cause X desire to be fulfilled. E.g.: the drink water because you are thirsty is rational because you are correct in thinking that drinking water will quench your thirst. You are not rational in brushing your hair because you are thirsty because brushing your hair will not quench your thirst.
- If you argue this case, you can give a subtle answer to the question in your conclusion – yes, but additionally the reason must be adequate, with X being the conditions for adequacy. Having a reason is a necessary condition for rationality – meaning you need one in order for your action to be rational – but it is not a sufficient condition – meaning it is not by itself enough to make an action rational.

➢ There are lots of things we do that we don't consciously justify or even choose to do, that we might still want to describe as rational. We might say something like being scared of heights is rational because it is a subconscious fear developed through the process of natural selection that stops us going near dangerous high cliffs, for instance, where we might fall and die. Though we can give an *explanation* of why are scared of heights, this isn't a *reason* because a reason is something you have for an action you choose to do. There are lots of examples of this – breathing, being around other people, sleeping.

Conclusion:
➢ Reassert your point of view and summarise the main arguments from your essay.
➢ Since most of this essay was concerned with definitions, restate your definition of reason and rationality, and summarise why that means a reason does or does not make your actions rational.

SECTION TWO — 2009

What changes in society will follow from increased life expectancy?

This question is very vague and leaves a lot of scope for you to decide exactly what you want to write about. Since the question does not specify things like by how much life expectancy will increase, or over what time frame this increase will occur, you should be sure to always fully justify and qualify your answer with under what conditions you believe these changes will occur – unless you believe they will occur under any conditions of increased life expectancy, in which case this, too, should be stated and justified. In terms of structuring the essay, you may wish to group your changes into categories – for example, legal changes, philosophical changes, lifestyle changes – in order to prevent your essay seeming like a list as you deal with each potential change.

Alternately, you could focus on the most significant changes that will occur, and explore the reasons for and against thinking they will happen in greater depth. In either case, be sure to explain the reasons one might have for believing these changes will not happen as well, and your reasons for dismissing these arguments.

Introduction:
- ➤ You should consider exactly what is meant by 'increased life expectancy'; presumably it amounts to an increase in the average age at which citizens die. There are two points to consider in relation to this:
  - Does it increase evenly across all demographics, or do the lives of the rich increase in length dramatically, while the lives of the poor not at all, or only by a relatively small margin?
  - How does your health fare as life continues?
- ➤ Doing this allows you structure your essay in the form "under X conditions, Y will occur, but under Z conditions, W will occur, for each condition to 'increased life expectancy' you consider relevant.
- ➤ You may wish to decide on one or a few of what you believe will be the major changes, and outline them here. Smaller points that will take up only a sentence or two later in the essay need not be mentioned in the introduction if you are covering a lot of changes, but be sure to outline your main points.

Potential changes:
- ➤ If health is still (in most of the population) good for the majority of one's lifespan, the ages at which people are expected (rightly or wrongly) to meet certain milestones will likely increase – this includes marriage, having children, and retiring.
- ➤ If health still declines severely after what we now consider to be middle-aged, healthcare services will have to expand rapidly to cover the rising demand.

We will also need to cover production of this, and other needs (including food production), with a proportionally smaller workforce, since less of the population will be of what is now considered 'working age'; our work will therefore have to become more efficient, and there may be a need to increase taxes to fund the necessary work.
- If access to the increased life expectancy is determined by, say, wealth, this may cause more fractioning and conflict in society, as those able to afford the healthy lifestyle that lends itself to longevity see even more tangible results.

Changes in the way we experience life:
- With more life, it is possible to argue either we will gain both in terms of quantity and quality of our experiences. As well as having more time in which to experience life, the extra time will lead to an increase mastery of skills by individuals (this will also depend on how healthy people are after a certain age). We may therefore see rapid development of technology or the arts as the masters in these fields have more time to work before passing.
- Alternately, with more life on our times, we may experience a decrease in a sense of urgency and motivation and so become complacent, and no such developments will be made, but instead development will slow as people no longer need to rush their cultivation of skills due to more limited time.
- We could also become bored with growing lifespans; this is especially the case if health still declines after middle age. With an aging population in ill health, society will have to develop ways in which to keep the population happy and entertained.
- People sticking around for longer could have two potential effects:
  - With more time to learn and gain experience, people gain emotional and moral maturity. Society will benefit from the increase in a number of people with much more life experience; we will be able to take guidance from them in more wisely writing laws, etc.
  - People's emotional and moral maturity will not increase as a result of longer lives because people stay the same after they have been socialised as a young person (after the age of about 25, say), and so increased life expectancy will have negative impacts on the moral growth of society as a whole; with slower turnover rates between generations, we will be 'stuck in our ways' longer, which will prevent positive social change.
  - Further to both these points, an increase in the elderly proportion of the population gives them more voting power, so whichever effect occurs, it will be amplified.

> Retirement is a major point that spans many of those above: with life expectancies increasing, provided they are in good health, people will be expected for longer. This may cause resentment amongst younger generations towards people born before themselves who benefitted from the younger retirement age. It may also cause conflict between classes, as the working class is forced to work for longer whereas those who can afford to retire early simply enjoy a longer retirement as a result of increasing lifespans.

Conclusion:
> Summarise the most important changes, and the most important variables in determining how society will change.
> It is fine to acknowledge that predicting such effects is difficult, and so not argue too strongly that any particular changes will definitely happen; acknowledge that the future is not certain, and make it clear which changes you think are more likely.

END OF PAPER

2010

Section 1

Question 1: E
The passage describes how most plastic bottles in the USA are PET, and how there are health risks associated with PET recycling. It then goes on to describe how PET does not take up much space in landfill, and how it is not harmful to the environment. At no point is it claimed that PET *cannot* be recycled safely, and this does not necessarily follow on from the passage, as there may be ways to evade the health risks. Thus A) is not a valid conclusion. D) is not claimed by the argument, and is not required for its conclusion to be valid, so D) is an irrelevant point, and is not a conclusion.

E) correctly identifies the main conclusion of the argument, as if all the reasons given in the passage are true, they give us good cause to believe the statement in E). B) and C) are both reasons given in the passage to support this conclusion, and are not conclusions in themselves.

Question 2: D
First we need to work out the ordinary cost for 2 adults and 4 children to go swimming 40 times:
- The 2 adults pay £2 each per session. This is a total price of £80 each, so £160 in total
- The 4 children pay £1 each per session. This is a total price of £40 each, so £160 in total

Thus, the total price without the swimcard would be £320.

The swimcard costs £50 for the year, and allows unlimited swimming for 2 adults and 3 children. The 4th child is charged at half the normal rate. This is 50p per session, so £20 in total for 40 sessions. Thus, with the swimcard the family paid £70 for their 40 swimming sessions.

£320-£70=£250. Thus the family saved £250 with their swimcard.

SECTION ONE 2010

Question 3: C
The passage discusses how there are many golf accidents, but not many golfers have insurance. It then concludes that insurance could reduce accidents. C) correctly points out that lack of insurance does not *cause* golf accidents, so the argument's conclusion is not valid.
E) is a completely irrelevant statement, so is not a flaw.
A) and C) are not flaws, because the argument is only claiming that insurance would *reduce* the number of accidents. Thus, the statements in A) and C) are irrelevant.
B) is also irrelevant for this reason. If some insurers did not provide adequate cover, this does not mean there would be no reduction in cases thanks to the other insurance policies.

Question 4: D
The argument states that owning handguns was banned in 1997, and that there has been an increase in crimes involving firearms since this action was taken. This does **not** mean that the law has *caused* the increase in crimes involving firearms, so A) is incorrect.
The passage also says that crimes involving legally owned firearms has increased by 0.1%, whilst the total increase has been 5%. However, none of this provides enough information to reliably conclude the statements in B), C) and E). We do not have any information on the number of legally owned firearms, so B) cannot be concluded. We have not been told anything about the number of crimes involving handguns, so C) cannot be concluded.
Equally, the fact that there are clearly now more cases involving illegally owned firearms does **not** mean it has become easier to acquire them (more people could have decided to go to the effort involved, or the number owned may not have changed, and there simply could be more *usage* of illegally owned firearms in crimes).
We can, however, reliably conclude that the law of 1997 has failed to reduce gun crime levels. Thus, D) is the answer.

Question 5: A
The passage concludes that we should not introduce payment schemes where people pay based on time spent on the roads. The reasons given to support this are based on studies done with driving simulators. At no point is it stated that driving simulators are a reliable guide to real life driving, and if this is not true, the argument's conclusion no longer follow on from its reasoning. Thus, A) correctly identifies an assumption.

SECTION ONE   2010

D) and E) are irrelevant as the argument does not refer to whether these schemes would effectively solve congestion or whether other schemes may have success. It simply says the dangers involved in these schemes mean we should not use them.
C) is also irrelevant, whilst B) actually weakens the argument's conclusion. Thus, neither of these are assumptions.

Question 6: D
The best way to answer this question is to simply find the lowest common multiple (LCM) of 6, 8 and 18. This will be the number of months in which the companies will all make a payment in the same month.
The LCM of these numbers is 72: ($6 \times 12 = 9 \times 8 = 4 \times 18$)
72 months is exactly 6 years. Thus, the companies will all make a payment in the same month in 6 years time, which will be January 1996.
Thus, the answer is D)

Question 7: E
We know the woman is not more than 12 hours late, so the latest time the clock can show is 19:00. This means the first digit on the clock must be either a 1 or a 2. The first digit on the clock cannot be a 1 as the top horizontal line would not be present. Thus, the time must be earlier than 10:00am.
Thus, we know the time must be between 07:00 and 09:59. Thus, the latest the time can possibly be is 09:59, meaning the latest the woman can possibly be from examining the first 2 digits of the clock is 2hr 59 minutes.
Now we look at the other two digits and see if a time of 09:59 is possible. The third digit could be a 5, so a time of 09:5X is possible (where X is an unknown number).
The fourth digit could be a 9. Thus, a time of 09:59 is possible. Thus, the latest the woman can be is 2hr 59 minutes.
Hence, the answer is E)

Question 8: B
We are told that an increase of 1 Unit on the Scale represents a ten-fold increase in the strength of the quake. Thus, we know that:
- Y is 10 times the strength of X
- Z is 10 times the strength of Y
- Thus, Z is 100 times the strength of X
- Thus, X is 1 hundredth the strength of Z

We can see that only statement B) fits in with these criteria.

SECTION ONE 2010

Question 9: C
The passage describes how demand for elephant tusks has fallen sharply. It then goes on to describe how *because of this*, there will soon be a decline in elephant poaching. This is the crucial point. Both B) and C) weaken the argument's conclusion, suggesting that high numbers of elephants may be killed. However, B) is not relevant to the argument's reasoning, as the killing methods do not affect how a decrease in demand will affect elephant poaching.
C), however, directly affects how a decrease in demand will affect poaching. It suggests that the decrease in demand *will not* cause a decrease in poaching, because poachers will stockpile for future rises in demand. This directly weakens how the argument's reasons lead on to its conclusion, so C) most weakens the answer. Hence, the answer is C).
A), D) and E) all strengthen the idea that elephant poaching is decreasing/will soon decrease. Thus, none of these weaken the argument.

Question 10: B
The argument describes how breast cancer cases, and average consumption of alcohol by women have both risen in the same time period. It concludes that this means the increased alcohol consumption is responsible for the increase in breast cancer cases. B) correctly points out the readily apparent flaw in this reasoning, namely that it has confused cause and correlation.
D) is not claimed by the argument at any point, whilst E) is incorrect as the argument is only presenting an explanation for an increase, so the size of the increase is irrelevant.
A) and C) are not flaws, because the argument specifically refers to an increase in the *average amount* of alcohol consumed by women in this time period, and the study referred to a number of regional hospitals, so is not overly-localised.

Question 11: C
The argument discusses how in the 18th century there was no way of killing weeds, and close-cutting lawnmowers did not exist. It the finishes with the statement: "given this, lawns must have been like flowery meadows". We can see that this last statement clearly identifies the main conclusion of the passage, which is the statement given in C).

D) identifies an assumption in the passage, claiming that a lawn will more readily exist in a state like a flowery meadow. This must be true in order for the conclusion to be valid, as if it is not, the fact that lawnmowers and weed-killers were not around *does not* necessarily mean that 18th century lawns must have been like flowery meadows.

B) and E) are completely irrelevant statements, whilst A) is an incorrect statement that cannot be concluded from the passage (the fact that without these technologies, the lawns were like flowery meadows does not mean that these are the *only* things keeping modern lawns smooth).

Question 12: D
First, we must calculate the cost of items costing £96, with the reduction.
- The reduction is 33%, of the total price.
- A third of £90 is £30. 33% is 1 hundredth less than 1 third. 1 hundredth of £30 is £0.30. Thus the total reduction of £90 is £30-£0.30, which is £29.70
- A third of £6 is £2. Again, 33% is 1 hundredth less than 1 third. 1 hundredth of £2 is £0.02. Thus, the total reduction of £6 is £2-£0.02, which is £1.98
- Thus, the total reduction would be £1.98+£29.70, which is £31.68. Thus, the price after the reduction would be £64.32.

Next we calculate the cost, with a reduction, if extra items are added to £6:
- £96+£6=£102. The total cost is now £102.
- 50% of £102 is £51. Thus, there is a £51 reduction.
- Thus the price after the reduction is £51.

£64.32-£51=£13.32
Hence, the answer to the nearest pound is £13. Thus, D) is the answer.

Question 13: E
The table appears confusing, but to answer this question we actually only need to consider the prices the house holder pair for standard and off peak electricity. We do not need to consider the total number of units *or* the price per unit.
First, we calculate the price she expects to pay for standard electricity for the next quarter:
- Last quarter she was charged £168.15 for standard electrical units used.
- We are told she has reduced her standard consumption by half. Thus, we expect the charge for this in the next quarter to be half of £168.15, which is £84.075 (or £84.08 to the nearest whole pence)

Then, we can calculate how much she expects to pay for off-peak electricity.
- Last quarter she was charged £10.06 for off-peak electricity.
- We are told she has increased her off peak consumption by around a factor of 10. Thus, we expect the new charge for off-peak electricity for the next quarter to be 10 x £10.06, which is £100.60

Thus, we expect her to be charged £84.08 + £100.60 for electricity consumed in the next quarter, which is £184.68. We then need to add on the fixed charges described in the last bill (£12.30 + £5.45), which will also be present. Thus, we expect a total bill of £202.43. Thus, we expect a bill of *around* £200.

SECTION ONE 2010

Question 14: A
We can readily see that if shape B) is used to fill the window, we would need to use 2 of the, one vertical and one horizontal, and that they would overlap in the middle, so B) cannot be used.
If shape C) were to be used, there would be a square shaped gap in the centre of the star shape, and shape C) would only fill in the points of the star. The same problem is encountered with Shape D).

Shape E) would cover this gap, but then each shape would overlap with the shapes either side of it, and thus E) cannot be used.
Shape B) would readily fill the window with no gaps left and no overlap. Thus, A) is the answer.

Question 15: B
A) cannot be confirmed or refuted as we have no information on the tolerance of smoking of young people as a whole. We do not know what percentages of young people approve of smoking, as we have only been given information on those that *do* smoke. D) similarly cannot be refuted or confirmed, as we have no information on how smoking affects a young person's relationship with their parents.
We can see from the table that 63% of parents of young smokers don't like the fact their children smoke, so C) can be dismissed as incorrect.
B) we can reliably conclude as correct. We know that only 19% of young smokers believe their parents do not like the fact they smoke, when in fact 63% of parents do not like it. Thus, B) can be reliably concluded.

E) cannot be concluded because we do not have information on other habits, or how likely parents of non-smokers would be to express strong disapproval of their children smoking. We can conclude that many parents of young people who smoke do not express their strong disapproval, but this does not necessarily mean that E) is true.

Question 16: E
The argument describes how RFR emissions of mobile phone masts have been frequently monitored for masts located near schools and hospitals, and that the levels have consistently been found to be below guideline levels.
The argument then concludes that there is no danger from Mobile phone masts. At no point has it been stated that RFR emissions are the only health risk from mobile phone masts, and we can see that if this is not true then the argument's conclusion no longer necessarily follows on from its reasoning. Thus, E) correctly identifies an assumption in the passage.

B) and D) are irrelevant as they do not affect the argument's conclusion that mobile phone masts do not pose any health threat. A) is also irrelevant as the fact that RFR emissions are more injurious to young people is irrelevant to the fact that the levels being below guideline levels means the mobile phone masts do not pose a health threat.

C) is also irrelevant, and would actually weaken the argument's conclusion, if true, so is not an assumption (an assumption must be needed for the argument's conclusion to be valid, and so cannot weaken an argument's conclusion if true).

Question 17: B
The passage describes how advertising raises costs involved in making products, and thus raises prices for the consumer. It concludes that this means that advertising disadvantages the consumer. If we accept the statement in B) as true, then this suggests that advertising actually benefits the consumer by resulting in a higher quality of goods, thus weakening the argument's conclusion.

A) does not affect the conclusion as the manufacturers' beliefs may be different from the truth, whilst D) is irrelevant as the passage is describing harm to the consumer, so a benefit to the unemployed is irrelevant.

C) and E) both reinforce the argument's conclusion by suggesting there is little benefit to the consumer from advertising, reinforcing the notion that advertising's effects may be negative on the whole.

Question 18: D
The person in the question weighs 80kg. Thus, in each journey in the lift, they can take 320kg of weight in boxes with them.
There are boxes weighing 25kg and boxes weighing 20kg. 12X25=300, thus they can take 12 boxes weighing 25kg on one journey. This leaves an extra 20kg of space, so they can carry one box weighing 20kg on each of these journeys as well. 120/12=10. Thus, in 10 journeys the person can transport all of the boxes weighing 25kg and 10 of the boxes weighing 20kg.

Now we need to account for the remaining 80 boxes weighing 20kg. 320/20=16, so 16 boxes weighing 20kg can be carried on a single journey. 80/16=5. Thus, 5 journeys can transport the remaining boxes weighing 20kg.
5+10=15. Thus, the minimum number of journeys is 15. Thus, the answer is D)

SECTION ONE — 2010

Question 19: B
The meeting is due to last at least 3.5 hours, thus this is the minimum length of parking required. The question has told us that the person cannot leave any later than 5:30, so 5.5 hours is the maximum length of parking required. Thus, we are looking for the car park which offers the cheapest parking for between 3.5 and 5.5 hours:
- Grove Street will cost £6.00 for this time range, thanks to the flat rate
- Victoria Square will cost between £4 (for 3.5 hours) and £6 (for 5.5 hours)
- Central Park will cost between £4 (for 3.5 hours) and £8 (for 5.5 hours)
- Bonningtons will cost between £4.50 (for 3.5 hours) and £7.50 (for 5.5 hours)
- Grange Road will cost between £4 (for 3.5 hours) and £7.50 (for 5.5 hours)

Thus, the cheapest option is Victoria Square. We can see that no car park has a cheaper charge than Victoria Square for 3.5 hours, and no car park has a cheaper charge for 5.5 hours than Victoria Square. Thus, the cheapest option overall is Victoria Square.

Question 20: D
The taxi charges at a rate of £70 plus £10 per passenger after the first. Thus, the prices are as follows:

1 person = £70.	4 people = £100
2 people = £80	5 people = £110
3 people = £90	6 people = £120

Per person, these prices work out as:

1 person = £70 per person.	4 people = £25 per person
2 people = £40 per person	5 people = £22 per person
3 people = £30 per person	6 people = £20 per person

Only Bar Graph D) follows this pattern.

Question 21: C
The question is describing how more people are choosing to use private healthcare, and how in the same time period, the NHS has suffered a decline. The argument then concludes that this means that people choosing private healthcare has caused the decline in the NHS. Answer C) correctly points out the flaw in this reasoning, that it could be the other way round. It may be people are choosing private healthcare *because* of the NHS' decline, not that people choosing private healthcare is *causing* the decline in the NHS.

E) is not a flaw in the argument because the argument has made no reference to funds being diverted to private healthcare, so this is not relevant to the argument's conclusion.
A) is an irrelevant statement. B) and D) are simply alternative arguments that counteract the one given in the question, and are not flaws in the presented argument.

Question 22: C
The argument's reasoning can be summarised as "A cannot happen because C is so high. C is even higher for B, so B cannot happen either". In this instance, "C" is the cost, and "A" and "B" are Amrik purchasing the 2 software packages.
Only argument C) follows this pattern of reasoning. Here, "C" is the caffeine content of the drinks, and "A" and "B" are Amrik drinking the 2 drinks after 9pm.
A) is different, because in this instance, the garlic is simply *present* in the foods. This is not the same as the garlic content being *even higher* than in a food we know Amrik does not like.
B) is summarised as "A cannot happen because C is too high", but then differs, because "C" (the distance) is lower for the option "B" which is given, so "B" can happen.
D) can be summarised as "A cannot happen because C is too high" (with A being finishing the jigsaw puzzle and C being the patience required), but then incorrectly states that because "C" also factors in "B", this cannot happen either. This is incorrect, and different from the question, because the crossword may require *less* patience than the puzzle.
E) is summarised as "A is less than C, B is greater than C, so A is less than B". This is not the same as the reasoning in the question.
Hence, the answer is C)

Question 23: D
The principle in the argument is that those who use a service more should pay more than those who use it less, such that those who use it less do not subsidise those who use it more. D) illustrates the same principle, with the toll system ensuring that those who use the road more are charged more than those who use it less.
B) is the opposite to this principle, with everybody paying the same regardless of how much they consumed at the restaurant.
C) and E) relate to situations where services are provided to, or charges are adjusted for, different people based on need, not how much of a given service is used. Thus, they do not follow the principle described in the question.
A) is a completely irrelevant statement. Thus, the answer is D)

SECTION ONE — 2010

Question 24: B

It is easy to calculate the possible distributions of letters in this instance. Let the three intended letter recipients be termed X, Y and Z:
- If the letter intended for person X is sent to person Y, then the letter for person Y must be sent to person Z, and the letter for person Z be sent to person X.
- If the letter intended for person X is sent to person Z, then the letter for person Y must be sent to person X, and the letter for person X be sent to person Y.

This accounts for all possible ways of ensuring everybody receives a letter intended for someone else.

Question 25: B

We are told that a serious claim costs 10 times as much as a slight claim, and that a write-off costs 10 times as much as a serious claim. Thus, a write-off must cost 100 times as much as a slight claim. Thus, to calculate relative costs, we should ignore the grant total of claims. Instead, we should treat each slight claim as 1 unit, each serious claim as 10 units, and each write-off as 100 units. Then, the total number of units will determine the relative cost of the month.

To avoid the need to calculate costs for all 12 months, we should simply calculate the months provided as possible answers, in the interests of time:
- January has 2891 slight claims, 539 serious claims and 30 write-offs. Under our system, this gives a total of 11281 units.
- March has 2972 slight claims, 550 serious claims and 31 write-offs. Under the system, this gives a total of 11572 units.
- June has 3230 slight claims, 509 serious claims and 24 write-offs. Under our system, this gives a total of 10720 units.
- October has 3236 slight claims, 474 serious claims and 30 write-offs. Under our system, this gives a total of 10976 units.
- Finally, November has 2895 slight claims, 578 serious claims and 21 write-offs. Under our system, this gives a total of 10775 units.

Thus, we can see that March has the most units under our system.
Thus, June is the most expensive month and the answer is B)

SECTION ONE   2010

Question 26: D
The exact figures are mostly difficult to read on this graph, so the best we can do is to examine the bank balance over the course of the year and deduce the pattern in which the bank balance goes up and down. Doing this, we put together a list of whether each change is positive, negative, or no change, and we see that the order is:
- Negative, Positive, Negative, Negative, Positive, No change, No change, Positive, Negative, Positive, Negative, Positive.

Thus we are looking for a graph that reflects these changes. An important thing to note is that we must remember to look for the change in bank balance in the correct month. For example, a positive change between February and March would result in the correct graph showing a "positive change" *in* the month of March, not in February. We can see that this means whatever change is shown for the month of January (the first month shown) is irrelevant, as we have no information on whether this balance has raised or lowered from the previous month (which we cannot see).

We see that only Graph D) accurately reflects the order of changes on the account.

Question 27: C
The passage describes how all supermarkets now possess loyalty cards, and thus they are no longer an incentive to shop at a single supermarket. It then states supermarkets should abandon these cards, and describes them as costly. It then backs up this conclusion by stating that they cannot be the main reason people shop at a supermarket.

All these reasons, if accepted as true, together provide good cause for us to believe the statement given in C), which thus correctly identifies the main conclusion of the passage.

A) and E) are both simply facts stated in the argument, and are not conclusions.
B) and D) are both valid conclusions that can be drawn from the argument, backed up by reasoning within the argument. However, they both go on to support the statement in C), which is also a valid conclusion from the argument. Thus, B) and D) are intermediate conclusions in this argument, which go on to support the main conclusion, given in C).

SECTION ONE 2010

Question 28: E

The passage describes how the demand for blood donors is growing much faster than the population of 18-65 year olds which can provide blood donors. At no point does the argument provide reasons why the demand for blood donors is growing, so D) is incorrect. It also makes no reference to changes in the supply of blood donors, or changes in the rate of growth of the relevant populations that can supply blood donors. Thus A) and B) are incorrect.

The argument does describe how there is no medical substitute for human blood, but *does not* say there would be no need for such a substitute if we could find more blood donors. There may be other reasons why it would be desirable to find a substitute, so D) is incorrect.

We can only conclude the statement provided in answer E). If the demand is growing faster than the supply, and a huge research effort has not found a substitute, then we can conclude that there are *no signs* of the problem of increased demand disappearing (we cannot conclude that it will not disappear, but we can conclude there are *no signs* of it disappearing). Thus, E) is the answer.

Question 29: A

The passage describes how there is no evidence that rhinoceros horn has an effect on the nervous system. It then describes how in a clinical trial of people taking powdered rice and people taking rhinoceros horn, more people taking rice had an increase in sexual arousal.

The argument then concludes that Rhinoceros horn is not an aphrodisiac. However, it could be that Rhinoceros horn actually does have aphrodisiac properties, and that Rice simply has stronger aphrodisiac properties, which would explain the results of the clinical trial. If this is true, the argument's conclusion is no longer valid. At no point has it been stated that rice is not an aphrodisiac, so this is a valid assumption in the argument. Thus, the answer is A).

B) and E) are irrelevant to the argument, so are not assumptions.
D) would actually strengthen the argument, adding further reasons why the claims of aphrodisiac properties may be false, so D) is not an assumption.

C) is not an assumption because it is not *required* to be true for the argument's conclusion to be valid. Powdered rice could have non-aphrodisiac effects on the nervous system, and the argument's conclusion would still be valid

SECTION ONE — 2010

Question 30: B

The simplest way to solve this question is to group the men into pairings, and see how much is owed by each person to each of the others.

Thus:
- Bill *owes* £200 to Fred, and *is owed* £300 *by* Fred.
- Fred *owes* £100 to Joe, and *is owed* £150 *by* Joe.
- Joe *owes* £200 to Bill, and *is owed* £250 *by* Bill.

By finding the differences in these amounts we can work out how much each person owes in total.

We see that:
- Bill is owed £100 by Fred.
- Fred is owed £50 by Joe.
- Joe is owed £50 by Bill.

Thus, Fred needs to give £100 to Bill. Bill needs to give £50 to Joe, and Joe needs to give £50 to Fred.

We see here that if Fred gives £50 to Bill, all the debts are cancelled. This is because Fred now owes Bill, Bill owes Joe, and Joe owes Fred, all by the same amount (£50). This is a pointless transaction, as this £50 will simply move around the group, back to Fred.

Question 31: C

We are looking for the cheapest way to purchase aftershave, Talc, body spray and 2 lots of shower gel. We can see that purchasing any pack offers a saving over purchasing all these items separately, so we should purchase as many items in packs as possible without spending extra money on unnecessary items.

We can see that the shower gel and body spray offers a discount for 2 of these products, without purchasing any unnecessary items, and that there are no other combinations which account for both lots of shower gel required without acquiring unnecessary items. Thus, we should purchase the shower gel and Body spray pack, costing £2.99

Now we have accounted for the body spray and one of the shower gels. Thus we can ignore the body spray double pack + free talc, as this purchases unnecessary items, which unnecessarily increases the cost.

There is now a choice to make. Either the Aftershave and Talc pack can be purchased, and then the other shower gel purchased separately, or the Aftershave and shower gel pack can be purchased, and the talc purchased separately:
- ➤ The aftershave and talc pack costs £4.49, and the shower gel costs £1.49. Thus, these purchases will cost a total of £5.98
- ➤ The aftershave and shower gel pack costs £3.49, and the talc costs £1.99. Thus, these purchases will cost a total of £5.48. Thus, this is the cheapest option.

Therefore, the cheapest option will cost £5.48, plus the £2.99 for the shower gel and body spray pack. This will cost a total of £8.47.

Thus the answer is C)

SECTION ONE — 2010

Question 32: C

We can see that in 1990, 40% of the adult population smoked, and in 1991, 30% of the adult population smoked. Thus, 10% of the adult population have stopped smoking between 1990 and 1991.

10% is 25% of 40%. Thus, there has been a 25% drop in the number of people who smoke if the number of people who have stopped is expressed *as a percentage of the number of smokers in 1990*. Thus, this is how the campaign organisers have expressed the drop.

Thus, the answer is C)

Question 33: D

The argument discusses how attempts to stop the supply of illegal drugs have not stopped drugs being readily available. It then goes on to say the USA should switch focus and instead attempt to reduce the usage of drugs, and suggests giving treatment to heavy drug users in prisons. If heavy drug users are in fact responsible for the majority of drug offences, then this supports the notion that this action will be effective in reducing the number of people using drugs. Thus, D) readily strengthens the argument.

A) and E) are irrelevant as they do not affect the argument's reasoning or its conclusion.

B) and C), meanwhile, actually weaken the argument. B) suggests that efforts to reduce drug supply are being successful, whilst C) suggests that the suggested action to reduce usage will not be successful, as the users do not wish to receive treatment.

SECTION ONE — 2010

Question 34: D
The passage describes how athletes/those with high levels of fitness are highly susceptible to disease, including cancer, placing them at risk of early death. It then concludes that in order to have a better chance of long-life, strenuous exercise should be avoided. D) correctly identifies the flaw in this reasoning. A high level of fitness may save more lives than the early cancer in this group kills. This would suggest that a high level of fitness *increases* your chances of a long life, thus making the passage's conclusion invalid.

B) is not a flaw because the fact that other reasons may contribute does not necessarily mean that a high level of fitness does not contribute, so the argument's conclusion may still be valid.

C) is incorrect because the passage describes the *chances* of a long life, so the fact that it might not work out that way is irrelevant to the conclusion.

E) is a completely irrelevant statement, whilst A) is incorrect because the argument has stated that studies have shown strenuous exercise damages the immune system, so there *is* evidence.

Question 35: D
The passage describes how road bumps should be abolished in residential areas. It backs this up with reasoning about how road bumps delay emergency services, and cause deaths from heart failures. It states how electronic speed signs and more effective speed limit signings would be just as effective at preventing road traffic accidents.

All of this, if treated as true, supports the notion that road bumps should be abolished in residential areas, which is the main conclusion of the passage.

E) and C) are reasons given in the passage to support this conclusion, whilst A) and B) are simply statements explaining why road bumps are there, a position which the argument goes on to suggest be abolished. Thus, none of these answers are conclusions from this passage. Hence, the answer is D).

Question 36: B
First we need to work out Ted's fuel consumption:
- ➤ 20km is travelled at 90km/h, as seen in the question. We can see from the table that travelling at 90km/h consumes 6 litres per 100km. 20km is 1/5 of 100, so this will consume 1/5 of 6 litres. Thus, this will consume 1.2 litres.
- ➤ 40km is travelled at 110km/h. Using similar calculations, we can see that this will consume 2.6 litres of fuel.
- ➤ 20km is travelled at 50km/h. Again using similar calculations, we can see that this will consume 1.5 litres of fuel.

Thus a total of 5.3 litres of fuel have been consumed. Fuel costs 80c per litre, so this will cost a grand total of 424c, which is $4.24. Thus, B) is the answer.

Question 37: B
First we must calculate how many sheets of paper each pack will use:
- ➤ The basic information sheet uses 1 sheet of paper, with both sides printed, giving 2 sides.
- ➤ The prices sheet also involves printing on both sides of 1 A4 sheet of paper, giving a further 2 sides.
- ➤ The information about local amenities involves 1 side of A4.

Thus, a total of 5 sides of A4 need to be printed for each pack (the colour brochure is provided for free by the local angling association so need not be considered). With 300 packs to be printed, a grand total of 1200 (5X300) sides of A4 need to be printed at the local printer.

Since the cost of printing is 2p per side, this means the total cost will be 2400p, which is £24.

Question 38: E
If the shapes are examined, we see that they all correspond to the orientations of the different symbols as shown in the 3 views in the question, apart from the die in E).

The Die in E) shows arrow pointing away from the X, which is the correct orientation, as seen in the left-hand view of the question. However, the crescent in Die E) is in the wrong orientation. We can see in the question that the Arrow points toward the face with the two parallel lines, which both lie perpendicular to the direction of the arrow. We can also see that one end of the 2 parallel lines point towards the face with the crescent on, such that they point towards the *concave* side of the crescent.

Thus, we can see that one of the tips of the crescent must point towards the face with the arrow on. In Die E), it is the concave side of the crescent that faces toward the arrow. Thus, Die E) is not the same as the one in the question. Hence, the answer is E).

SECTION ONE — 2010

Question 39: B

The passage describes how when speed cameras are painted yellow, drivers drive quickly between them, braking when they see one. The argument claims this is a dangerous practice. It then describes how if the cameras were almost invisible, drivers would have to stick to the speed limit all the time, for fear of being caught unawares. This does *not* mean that speed cameras were difficult to see, drivers would be *likely* to keep within speed limits. The drivers may simply disregard the risk of being caught by a speed camera, so this conclusion does not necessarily follow on from the passage. Thus, E) is incorrect.

However, the reasons given *do* suggest that making speed cameras visible is not a sensible policy, as it claims the response to obvious cameras is dangerous, which readily gives this conclusion. Thus, B) is the answer.

A) is incorrect as at no point does the argument *refute* the suggestion that obvious cameras act as a deterrent to speeding motorists. It merely claims that this notion is the opposite of what should be followed.

D) is a summary of the reasons in the passage leading onto the conclusion in B), whilst C) could be described as an assumption in the passage, which invalidates the conclusion if it is untrue. Thus, neither of these are conclusions from the passage.

Question 40: E

The passage describes how success in America is measured by possessions, and that those with few possessions are *judged* to be unsuccessful. It then concludes that those with few possessions must feel a strong sense of failure.

For this conclusion to be valid, we are required to accept as true the idea that those with few possessions *wish to be seen* as successful. If this is not true, then the fact that others judge them to be unsuccessful is irrelevant, and does not necessarily lead on to the notion that they will feel a sense of failure. At no point is this idea stated, it is simply assumed. Thus, E) correctly identifies an assumption in the passage. None of the other answers are relevant to the argument's conclusion, so none are assumptions.

SECTION ONE — 2010

Question 41: C

The passage describes how different countries in Eastern Europe (such as Greece, Cyprus, Russia and Ukraine) are consistently voting for one another. It then discusses how as more eastern European countries enter the competition; Western nations will find it difficult to win, before concluding that the result is now decided by geo-politics, not music.

However, if it is true that Eastern European countries have similar tastes in music, this suggests that the Easter European countries' habit of voting for each other *is* based on music, not geo-politics. Thus, C) weakens the argument's conclusion.

B) is a completely irrelevant statement, whilst both D) and E) refer to *past* competitions, so bear no effect on conclusion, which refers to the result *now* being decided by politics, in the present tense. Thus, both D) and E) are irrelevant

A) actually strengthens the argument by implying that western powers have their own, geo-politically decided voting blocks, strengthening the conclusion that geo-politics, and not music, decides the competition.

Question 42: D

Firstly we see that there are 2 possible coffee varieties, caffeinated or de-caffeinated.

Next, we see that each of these 2 possible varieties allows black, or 3 sorts of milk, giving a total of 4 options. Thus we now have a total of 8 options (2X4=8).

Next, we see that for each of these 8 options, we have a total of 5 options relating to sweetener (those being no sugar/sweetener, 1 spoonful of sugar, 2 spoonfuls of sugar, 1 sweetener or 2 sweetener).

Thus, this gives a total of 40 possible options (8X5=40).

Question 43: A

As the question explains, each team carries through half their points from the first round to the next. At the end of this round, the winner subtracts the points carried through from their first round from their total score, to work out their "new" points. This number is then halved, and the resultant number carried through to the next round.

Thus to answer this question, we must track each team from the first round, applying these calculations as described in the question, to see how many points each team will enter the final with.

First we shall assess the Indigo Iguanas:
- The team earned 82 points in the first round, so would have carried through 41 for the Quarter Finals.
- They finished the quarter finals with 125 points. 125-41=84, so the team earned 84 new points in the Quarter finals. Thus, they carried through 42 to the semi-finals
- They finished the semi-finals with 118. 118-42=76, so the team earned 76 points in the Semi-Finals. Thus, they will carry through 38 points to the final.

Now we assess the Orange Ocelots:
- The team earned 76 points in the first round, so carried through 38 points through to the Quarter-Finals.
- They finished the quarter-finals with 130 points. 130-38=92, so the team earned 92 new points in the quarter-finals. Thus, they carried through 46 points to the Semi-Finals.
- They finished the semi-finals with 128. 128-46=82, so the team will carry through 41 points into the final.

Thus, the Indigo Iguanas will begin the final on 38 points, and the Orange Ocelots will begin on 41. Thus, the Indigo Iguanas must earn 4 more points than the Orange Ocelots in the final in order to be victorious.

Thus, the Answer is A)

SECTION ONE 2010

Question 44: C

There are 5 starred points identified. One of these occupies a central location where 6 paths converge on one point. 3 of the other starred locations are to the left of this point, whilst one is to the right of this central point. Let the 3 starred locations to the left be termed A, B and C, the central point be termed X, and the starred point to the right be termed Z.

We can quickly see that to go from *any* of points A, B or C to point Z, or vice versa, will *always* require going past point X. We can see that X is the *only* point which will *always* have to be passed to make a certain journey. Thus, we can identify X as the Bandstand, which *always* must be passed to move from the statue to the fountain. Hence, we can also see that the Fountain and the statue must be on opposite sides of the bandstand, otherwise you would not *always* have to pass the Bandstand to move between them.

Thus, we now know that the one out of the Fountain and the Statue is Z, and the other is one of A, B or C.

We are then told that to move from the lake to the bowling green, the fountain is *never* passed unless the bandstand is passed twice. This means that the Bowling Green and Lake must both be located on one side of the Bandstand, and the Fountain on the other (otherwise the bandstand would not *always* be passed twice whenever the fountain is passed). Since there is only one point on the right-hand side of the Bandstand, the Bowling Green and the Lake cannot *both* be located there. Thus, the fountain *must* be point Z in our labelling.

Thus, we can identify Z as the fountain, and X as the Bandstand. We know that A, B and C must be the other 3 monuments, but we cannot determine which ones they are.

Thus, 2 points can be identified from this information.

Question 45: B

The passage discusses how people take care of passports better when they are a book-style document which cannot be easily slipped into a pocket, and then easily lost. Ordinarily, we would not assume that this necessarily gives a conclusion that the proposal to replace book-style passports with small plastic cards should be rejected.

However, the passage has stated that the proposal should be rejected, near the start. This means that the passage has identified this notion as its conclusion, and is providing reasons to back up the conclusion. Thus, B) correctly identifies the main conclusion of the passage.

A), C) and D) are all intermediate conclusions from this passage, which all go on to support the main conclusion given in B). Thus, none of these are in themselves the main conclusion.

E) is not a valid conclusion from the passage. The passage actually states that the proposal should be rejected outright. This is different to suggesting that the proposal be examined more carefully, so E) is incorrect. Hence, the answer is B)

Question 46: D

The passage's reasoning can be summarised as "A must happen for B to happen. A will not happen, so B cannot happen". Here, "A" is Nusra accepting a lower salary, and "B" is her finding an interesting job. Answer D) follows this same reasoning, where "A" is Paul and Lisa talking to each other more, and "B" is them having a happy marriage.

B) and C) both identify that "A must happen for B to happen", and point out that "A" is not currently happening. However, they both then go on to suggest ways in which "A" can be achieved (e.g. by Julie and Mike working hard at stability, or by Myra borrowing the rest of the money), which is not the same as simply concluding that "B" cannot happen.

A) also follows different reasoning, and can be summarised as "A must happen for B to happen. We want B to happen, so A must happen". This is different from the reasoning given in the passage.

E) can be summarised as "A must happen for B to happen. B is identified as the only option, so trying to stop A is pointless". This is not the same as the passage's reasoning.

SECTION ONE — 2010

Question 47: C

The passage discusses a principle in which rather than educating people about the dangers of something which can hurt them, we should simply ban the hurtful thing.
- C) follows the same principle, arguing that rather than educate about the dangers of alcohol, we should simply ban alcohol.
- A) follows an opposite principle, whereby rather than banning a hurtful thing, we should educate people about the dangers of it.
- B), D) and E) all discuss the effectiveness of different methods employed to attempt to educate people about the dangers of a certain thing. Thus, none of them have any reference to banning the hurtful thing, and thus none follow the principle in the question.

Question 48: B

The Hovercraft takes 45 minutes to travel between ports, and stays in berth for 15 minutes at each Port before travelling to the other. Thus, if the Hovercraft leaves at 8:00am, it will arrive at the next port at 8:45am, and depart at 9am. It will then arrive at the other Port at 9:45am, and depart at 10:00am.

We can see here that the Hovercraft will always be in Berth between 45 minutes past an hour, and the start of the next hour. It will be in each of the 2 ports as follows:
- 08:45-9:00 – Selmar
- 09:45-10:00 – Harport
- 10:45-11:00 – Selmar
- 11:45-12:00 – Harport
- 12:45-13:00 – Selmar
- 13:45-14:00 – Harport
- 14:45-15:00 – Selmar

We now need to calculate the berthing times for the Ferry, and see when they match up with the Hovercraft. The ferry takes 75 minutes to complete its first journey to Selmar, thus arriving at 9:15. It then waits 30 minutes before departing for Harport at 9:45.

Thus it will arrive at Harport at 11:00. It will then depart Harport at 11:30, and arrive at Selmar at 12:45. We can see from the list above that the Hovercraft will also be at Selmar at 12:45.

Question 49: D

We know that the telephone numbers are all 6 digits long and that they all start with "58". The first thing to notice about the 4 digits given in the paper is that they all look the same upside down. Thus, this piece of paper could read "8011" or it could read "1108".

Thus, we have 3 possibilities about what these numbers can represent.
- One possibility is that the first 2 digits have been ripped off, and that these 4 digits read "8011", and represent the last 4 digits of the phone number. This gives us 1 possibility for what the phone number could be, as we would therefore know all the digits
- Another possibility is that the first 2 digits have been ripped off, and that these 4 digits read "1108", and represent the last 4 digits of the phone number. This gives us 1 more possibility for what the phone number could be, as we would therefore know all the digits.
- The other possibility is that the first and the last digits have been torn off, and that these digits read "8011" and represent digits number 2, 3, 4 and 5 of the phone number (thus the "8" would be part of the "58" that we know the number starts with). If this is the case, we know what the first digit must be (a 5), but we do not know the last digit. Thus, this would allow 10 different possibilities for what the number could be.

We know that the 4 digits cannot be the first 4 digits, because the number has to start with 58. Because of this, we can also see that the numbers cannot read "1108" and represent digits 2-5, as this would also mean the phone number didn't start with 58.

Thus, there are 12 different possibilities for what the phone number could be.

SECTION ONE 2010

Question 50: A

We can see from the right-hand view that the face with 6 dots is adjacent to the face with 1 dot, such that the lines of 3 dots run *towards* the face with 1 dot on, not perpendicular to it. Thus, in the second-left view, the face with 1 dot on could be to the right or the left of the face with 3 dots on. This means that the face with 1 dot on must also be adjacent to the face with 5 dots on.

We can also see from the left-hand and second-right views that the face with 3 dots on is adjacent to *both* the face with 1 dot on *and* the face with 4 dots on. We are looking at the face with 3 dots from the same direction, and can see that in these 2 pictures it has either been rotated 90° or 270°, depending on which direction we consider the rotation in. Thus, we know the face with 1 dot on must also be adjacent to the face with 4 dots which.

Thus, returning to the second-left view, we now know that 3 of the faces we cannot see are the faces with 3, 1 and 4 dots on. All of these faces are adjacent to each other. This means they must occupy the face at the rear side of the die (i.e. opposite the 5 dots), the bottom of the die (i.e. opposite the 6 dots) and *1* of the sides (i.e. to the left or to the right of the 5 dots).

This means that the other unseen side in the second-left view must be the face that we have not seen in any view, with 2 dots. Since we know that the face with 1 dot is on one of the sides in the second-left view, this means that this face *must* be opposite the face with 2 dots.

Thus the face opposite the face with 2 dots has 1 dot.

Thus, the answer is A)

END OF SECTION

Section 2

'Printing and the telephone were truly revolutionary inventions. All the internet brings is a difference in scale.' Is that true?

This essay can be structured well by doing so simply – begin by defining the terms of the argument (what constitutes a 'revolutionary' invention?), then laying out your side of the argument. Follow this, or follow each individual argument, with the counterargument another person might use, then your reasons for considering them incorrect. Conclude simply and concisely by restating your position and summarising your overall argument.

Introduction:
- Establish what you consider to be a 'revolutionary' invention; you may wish to define it as an invention that significantly changes the way in which we act in whatever field it is concerned with (in this case, information and communication).

- Outline the reasons that printing and the telephone were revolutionary inventions, then state your position on the internet, and outline the main arguments you will be using.

- If you have an overall theme to your arguments – for example, you might have argued that the internet has subsumed the functions of printing and the telephone, and so could be mistaken for a simple expansion in scale, but is still revolutionary due to the additional functions it provides – this should be stated clearly in the introduction.

Potential arguments for:
- It is worth first establishing that printing and the telephone were revolutionary inventions (it is unlikely you will want to argue against this, but you can do if you have sufficient justification for doing so). Both technologies vastly expanded our potential to communicate, reproduce, store, and transport information. They also represented large technological leaps.
- A lot of the functions the internet provides are the same as printing and the telephone, kicked up a notch. The two major groups into which these functions can be placed are communication and information storage. The telephone allowed us to communicate with people across the world, while the invention of printing allowed us to reproduce (at a much faster pace and with much less labour intensive work) and store information so that it can be accessed by anyone at a later time. The internet provides both of these services, but faster, and on a much larger scale; we can make information public so anyone with internet access can find it (for free, if we so choose), not only those to whom we send the printed word, or those we call up.
- The internet made the functions provided by printing and the telephone cheaper and accessible to many more people; nearly everyone in the developed world has access to the internet (whether in their home or in, say, a nearby public library), which allows far more people to publish work (online) than physical printing, an expensive process only accessible to a few through publishers, etc., ever did.

> The internet has taken over the functions of the telephone and printing – more and more people are using the internet to stay in touch with people close to them (email, Facebook, etc.), and to otherwise contact people (emailing employers with job applications rather than sending them letters or phoning them, for instance), and more work is being published online all the time (e-books, online newspapers, blogs).

Potential arguments against:
> Printing, the telephone, and the internet all have very different functions; while the internet combines some of the functions of the previous two, it is a technology that has not subsumed them and has functions they never provided. These functions include communicating instantly with potentially any number of people, providing a platform to share videos and audio media, searching through vast amounts of information to find what you need, and entertainment in various forms.
> That the internet has not subsumed the telephone or printing is demonstrated by the simple fact that both technologies are still widely in use.
>> **Counterargument:** but the roles of both are fulfilled by the internet (most mobile phones, for example, now allow you to access the internet, which is widely used for instant communication), and the use of telephones and printing is in decline; that they are still in use at all may just be indicative of old habits dying hard, but society as a whole is still moving away from them, and will eventually stop using them all together.
>> However, this does not necessarily mean that the internet is not revolutionary – it still provides functions that neither of the previous technologies did.

> You could argue that the difference in scale is itself revolutionary; giving the public at large access to the means of publishing their own work, finding information for free, and communicating with everyone from family members to their local politicians instantly vastly changes the way we interact with information and communicate with each other; this is a revolutionary change.
> This is especially the case for communication with individuals, not simply with stored information. Historically, communication between the powerful and other people ('powerful' including politically powerful people, celebrities, experts in various fields of study, etc.) has been very one directional; the powerful publish papers or give press releases, and the general public read it. For the first time in history, this is becoming more two directional – and the internet is to thank. People can do everything from send emails to MPs to tweeting Obama to sending reviews of published work straight to the creator, all at no cost to themselves, and quite often to recognition and response from the people they are contacting.

Conclusion:
> Restate your position and summarise the arguments you've presented above.
> Link your conclusion back to the question by referring to the concept of revolutionary inventions, and why the internet does or does not qualify.
> If you have a more nuanced argument, such as the example given in the introduction, this should be restated here with the arguments that justify it.

SECTION TWO — 2010

Is it justified to insist on facial visibility in public spaces?

This question is easy to structure well. In your introduction, clearly state your view and outline the arguments you will use to support it, as well as potential counterarguments and your reasons for dismissing them. It is important to note the vague wording of the question; the interpretation of 'public space' is left open, and it is particularly important to give a clear definition of what a public space is. Regardless of the definition you give, it will most likely be a broad category, so it is fine to give a nuanced answer that outlines the criteria for in which public spaces it is justified to insist on facial visibility, and in which it is not.

Introduction:
- It is important here to establish what you take 'facial visibility' and 'public spaces' to mean. Facial visibility might be defined as having your face recognizable as you; this means no clothing covering it, and not so much makeup that you become unrecognizable. 'Public spaces' is trickier to define, but we might do so by saying it encompasses spaces which are free for anyone (or anyone of certain demographics, for example, bars which only admit adults) to enter – this includes shared public land, e.g.: streets and parks, and private land that has been opened to public entry, e.g.: shops.
- Once you've defined the terms of the question, clearly state your position and outline your main arguments, the counterarguments to these, and your reasons for dismissing them.

Potential arguments for:
- Most crimes which could be caught on CCTV occur in public places; insisting on facial visibility would improve our chances of identifying criminals after the fact, and discouraging people from committing these crimes in the first place.
- Further to the above point, the only people who would object to this reasoning and insist on covering their face are those who intend on committing crimes – only the guilty have something to hide!
- It's fine if people don't want to be seen or recognised – but if that's the case, then they should stay home; having unidentifiable people out in the street is a potential danger.
- A more nuanced view might argue that facial visibility need not be insisted upon in all public places (and so is not acceptable, since it would only be acceptable to violate a person's right to self-expression and privacy if it were necessary for some greater good), but it is in some e.g. where it's *necessary* to verify someone's identity like at airport security or court.
  - **Counterargument:** this misses the very crux of the argument – the question does not ask whether there are some places where insisting on facial visibility is necessary and are some of these places, coincidentally, public places, but rather is asking whether there is something about the nature of public places that makes insisting on facial visibility acceptable; this says nothing to support the view that it does.

- ➤ Given the above counterargument, we might want to consider which public spaces our previous arguments regarding crime and identifying criminals apply to. Crime that could be discouraged or sanctioned by insisting on facial visibility might be prevalent in shops, but not by students in schools (where they would be identifiable by their clothing, mannerisms, etc. to the teachers who know them, if any misdeed were to occur at all). If we are insisting on facial visibility on this basis, it should only apply to those public places where it is beneficial.
  - o **Counterargument:** this is ridiculous; crime and other misdeeds occur everywhere at one time or another, but even if we were to narrow our insistence on visibility to those places where it is most prevalent, it would be deeply unfair to effectively ban people who wish to cover their face from these spaces, and to treat them as a criminal for this simple desire for privacy.

Potential arguments against:
- ➤ Insisting on facial visibility is discriminatory against people of some religions and cultures in which the face is traditionally covered.
- ➤ Facial visibility is only considered acceptable to insist upon in the UK because facial visibility is generally viewed as the norm. Disallowing baggy clothing that conceals the shape of one's bodies would also aid in, for example, recognition of criminals, but would be considered a ridiculous ban because of cultural norms. This is further exemplified when we look specifically at facial visibility – large sunglasses might be considered acceptable because they are not particularly unusual in dominant British culture, whereas a cloth face covering which does no less or more to obscure personal identity might not. Although facial visibility is considered the norm, or even better and more polite, by much of the UK, this is not an adequate justification for insisting on it at the expense of other people.
- ➤ People have a right to privacy, which includes privacy over their body.
- ➤ People can wish to avoid detection and recognition for reasons other than criminal activities – for example, avoiding someone who has been abusive towards them. It is important that people be able to avoid those who pose a threat to their safety.
  - o **Counterargument**: if someone poses a threat to another person's safety, our police force ought to be dealing with that; if they are not, the response is to improve our protection services and justice system, not to allow members of the public to hide from dangerous people better.
  - o **Counterargument**: additionally, allowing people to cover their faces also allows the dangerous person to conceal themselves, so we are no better off.

Conclusion
- ➤ Restate your position and summarise the arguments you have used to support it.
- ➤ Link your conclusion back to the question by referring to your definition of 'public space' and what about that, specifically, makes insisting on facial visibility acceptable.

SECTION TWO — 2010

Why do we need banks?

This question implies an assumption that we *do* need banks; it is perfectly fine to challenge this assumption and instead argue that we do not, as long as this position is well justified. In either case, you should provide some reasons that we don't need banks (either supported or refuted), and attempt to challenge the reasons you give for us needing banks (again, either supported or refuted).

Introduction:
- Give an outline of the functions banks perform, then clearly state your position on whether these functions are needed (or *which* of these functions are needed).
- Outline the arguments you will be using to support your position, and the counterarguments you will present and your reasons for dismissing them.

Reasons we need banks:
- Banks enable the flow of currency by providing a space in which to store it, and then transfer it instantly.
- In a globalised economy, dealing only with physical currency is completely unfeasible; everything from individual online purchases to dealings between massive multinational corporations would either be impossible or deeply impractical. Securely sending that amount of money between people could take days or even weeks in some cases, and would be simply not worth the effort and expenditure for most people.
- It is necessary to be able to safely store savings; though robberies (both physical burglaries and digital crime) of banks do occur, money is much safer than if people were keeping it in their homes, given that most people do not have the means to invest in a secure safe, whereas a bank account is not only free, but often comes with monetary rewards.
- Banks provide loans, which are necessary for investing in the future, for everything from buying a house that could otherwise only be afforded in retirement after a lifetime of saving, to starting up businesses before they begin to turn a profit. Banks make commodities and investments (in one's own business, education, etc.) accessible to more people.
- Banks are ingrained into our economy and provide basic services such as currency exchange, organising money into different accounts (savings accounts, current accounts, accounts for organisations managed by an individual, etc.), making online purchase, etc. Even if we once could conceive of an economy without banks, they are now far too integrated to ever consider doing away with them.

Reasons we don't need banks:
- While it is true that banks provide the function of storing and transferring money (which may well be necessary in today's economy), that is not their primary purpose; they exist to invest money and make a profit and this, as we have seen (think of the recent 2008 economic crisis) makes them dangerous and unstable. If it is storing and transferring money that is essential, we should be building more institutions like building societies, which are owned by their members and don't invest the money stored in them.
- This is similarly the case for loans; loans provided by banks are done for the sake of the bank wishing to make a profit from interest payments, not because they are interested in providing a start for small businesses. Their ultimate concern in profits means any positive economic outcomes are an accidental by-product, whereas they should be the focus of investment by our government and community.
- Loans are dangerous to the individual if they fail to succeed in repaying according to the agreed terms, and do little to combat inequality in opportunity, since they put those who cannot afford what they need out of their own pocket at the mercy of banks, while those who can do so can invest as they wish. We consequently ought to move towards more stable and community-orientated means of investing in small businesses, such as crowd funding.
- Though it's true that banks are heavily integrated into our economy, this doesn't mean we can't ever move away from relying on them, it just means it will have to be done slowly, and carefully until they are no longer necessary; if we believe banks are ultimately harmful and destabilising, then this needs to be done.

Conclusion:
- Restate your position and summarise the arguments you have used to defend it.
- Be sure to link your conclusion back to the question by referring to the roles of banks viewed as necessary (and whether you believe they actually are).

If two reasonable people claim the same fact as evidence for opposing conclusions, does it follow that it can't actually be evidence for either?

This is an interesting question with lots of room for debate on both sides, hinging on what we take the precise definition of 'evidence' to be. Given that it rests on giving quite a precise definition of quite a complex concept – evidence – it is good to make a comment on the complexity of the argument, and that is it unlikely we will be able to find an answer which *any* two reasonable people would themselves agree with. That said, it is fine to have a very strong opinion one way or the other, though of course be prepared to justify it, and give your reasons for dismissing other definitions.

Introduction:
- It is first important to establish what you mean by 'evidence'; "the available body of facts or information indicating whether a belief or proposition is true or valid" is one available definition.

- We should also note the use of the phrase 'reasonable people', and establish that we take this to mean that the conclusions drawn are logically consistent with the evidence, and are sound conclusions to draw on the basis of the piece of evidence under dispute alone.
- Clearly state your position, linking it to what you take to be the criteria for a fact being 'evidence' and outline the arguments you will use to defend it, as well as opposing views and your reasons for rejecting them.

Potential arguments for:

- If evidence is a fact that can be used to draw a reasonable conclusion, then the fact that an observation can be used to draw two reasonable conclusions shows it is useless for conclusion drawing. While it may be consistent with both conclusions, and be a relevant consideration (rather than, the observation "some dogs are brown" being *consistent* with the conclusions "the bus will arrive at 3:00" and "the bus will arrive at 4:00", but not *relevant* to them), it cannot be said to be 'evidence' because it doesn't help us draw a reasonable conclusion.
- Suppose that two facts, X and Y observed in isolation, both allow you to reasonably draw conclusions A and B but, when observed together, allow you to reasonably draw only conclusion B. You may argue that X and Y must both alone be evidence of B but this is not true; since they can be used to definitively establish B as the reasonable conclusion only when observed together, it is the conjunction of X & Y that is evidence of B. A and B alone are not evidence of X or Y- they are merely consistent with them.
- If two reasonable people can take a fact to be evidence of opposing conclusions, this is indicative of insufficient data to draw a conclusion at all – so we cannot say we have evidence either way. The conclusions drawn are most likely a result of bias from previously held opinions.

Potential arguments against:

- Since evidence is nothing more than a true observation that is indicative of some fact about the world, any true observation that is made as a result of some fact about the world is evidence of it. Take evolution, for example: if we rightly observe that giraffe's necks have lengthened over generations, we can take this as evidence of natural selection breeding giraffe's with longer necks, or we can take it as evidence of Lamarckism, which would say a giraffe's neck stretches during its life in order to, say, reach high leaves in trees, and this change is passed down genetically. Though without further information, a reasonable person could take this as evidence of either theory, it can only *actually* be evidence of a true theory; being taken as evidence of both does not negate that it is evidence of one.
- Alternately, taking the definition "the available body of facts or information indicating whether a belief or proposition is true or valid" we can argue that if an observation is consistent with some conclusion then it is evidence for it; if we have a set of observations used to justify two opposing conclusions, this is simply indicative of us having insufficient evidence to draw a reliable conclusion, but does not mean it is not evidence.

- ➢ We could further suppose that what is and is not evidence of a conclusion are not fixed categories, into which all facts definitively fall. In the case of the above example, could we say that the fact that giraffe's necks have elongated generationally is, alone, evidence of both natural selection and Lamarckism but, when combined with other observations, which are incompatible with Lamarckism, it is actually evidence of natural selection. Since 'evidence' is a function of human reasoning (being defined as that which we can use to draw a supported conclusion), rather than a category of thing that naturally exists, we can define it in terms of whether it *is* used to draw a conclusion, or whether it is *reasonably* used to draw a conclusion. In these situations, a fact was at one point evidence for both, and then evidence for only one.
- ➢ The way in which people use very difficult to accurately define terms such as 'evidence' is context dependent; what is 'evidence' in one situation may not be so in another. To make sweeping statements such as "if two reasonable people claim the same fact as evidence for opposing conclusions, it follows that it can't actually be evidence for either" is to ignore the nuance of language and communication.
- ➢ Being a 'reasonable person' does not necessarily mean your reasoning will always be flawless, and that you will always draw correct conclusions from a piece of evidence. Further, it seems absurd to suggest that the status of a fact as evidence for a conclusion is at least in part dependent on no one misinterpreting it. For example, let us say reasonable person A sees that things tend to fall to the ground when released and takes this as evidence of a force pulling us towards the earth. We would probably quite reasonably say their observation of facts were evidence of this force (let us also suppose it is true that gravity exists and was correctly observed here). Later, reasonable person B also observes this fact, and incorrectly concludes that this is evidence of a force pushing us down, away from the earth's atmosphere. Supposing this conclusion is correct, and opposing person A's correct conclusion – does the fact suddenly *stop* being evidence, where previously it was? This seems ridiculous, and yet is what the question is proposing.

Conclusion:
- ➢ Link your conclusion back to the question by referring to the definition of 'evidence', what does it mean for something to be 'evidence'?
- ➢ This is the sort of question where it is nice to make some concluding remark about the debate itself – does the precise definition of 'evidence' *matter*, so long as each person in a particular conversation is clear on what is meant by it? Does the definition of 'evidence' rigid and the same across all conversations and usages? Can something that is said to be evidence of a belief at one point stop being evidence of that belief when combined with other observations? Use this to show the examiner you are really thinking about and engaging with the implications of the question and wider field of study itself, not merely answering one question within it.

END OF PAPER

2011

Section 1

Question 1: A

C) is simply a stated reason stated in the passage, so is not a conclusion.

B) and A) are both valid conclusions from the passage, but only A) is the main conclusion. We can see that the statement in B) actually goes on to support the statement in A). Therefore, A) is the main conclusion and B) is an intermediate conclusion, supporting the main conclusion.

D) and E) are both statements which would weaken the argument, if true. Therefore, they are not conclusions. E) also appears to be incorrect, as it seems to have incorrectly concluded that because 39% of nurses working alternating shift patterns want to leave, the other 61% prefer alternating shift patterns. This is not a valid deduction.

Question 2: C

The simplest way to answer this question is to go through the answers and see if each one can be achieved from the possible scores for 4 matches:
- 8.5. points can be achieved through 2 draws, an away win and a home win (3+3+1+1.5=8.5), so A) is a possible score.
- 9 points can be achieved through 2 draws and 2 away wines (3+3+1.5+1.5=9), so B) is a possible score.
- 9.5 points cannot be achieved, so C) is not a possible score.
- 10 points can be achieved through 3 draws and a home win (3+3+3+1=10) so D) is a possible score.
- 10.5 points can be achieved through 3 draws and an away win (3+3+3+1.5=10.5) so E) is a possible score.

SECTION ONE 2011

Question 3: C
The argument has described how many low-budget films are very popular, and how many high-budget films are not successful. However, it does not follow that this is *always* the case, so the argument's conclusion is not valid. C) correctly points out the flaw in the argument's reasoning that it need not be cost that determines a film's popularity.
The rest of the possible answers are all irrelevant in that they do not affect how the argument's conclusion follows on from the given reasons. Therefore, the rest of the answers are not flaws in this argument.

Question 4: E
The argument describes how the only thing that can slow down global warming is an immediate switch to green technologies. It then discusses how if global warming continues *at this rate*, many species will not be able to reproduce. Thus it can be concluded that if we do not switch to green technologies, some species will not survive. Therefore E) is the main conclusion. It does *not* follow from this that the species *will* survive if we do switch to green technologies, it can only be said that they *won't* survive if we don't. Therefore C) is not a valid conclusion.
B) and D) both contradict statements in the passage, so are not conclusions.

A), meanwhile, is irrelevant, as nothing has been claimed about whether species cannot breed in warmer climates. It has only been said they cannot *survive*. Also, it has only been said that *some* species cannot survive, not *most*. Thus, A) is an irrelevant statement, and not a conclusion.

Question 5: B
The argument concludes that we should legalise Cannabis in order to reduce its use. It backs this up with reasoning describing how the dangers of Cannabis need to be more widely known if its usage is to decrease. At no point has it been stated that the dangers will not be widely known without legalising cannabis, and if this is not true the argument's conclusion is no longer valid. Therefore, B) is an assumption in this argument.
D) is irrelevant because the argument has stated that it is more dangerous than people realise, so we accept this as true. C) and A) are both irrelevant to the idea that legalising cannabis would allow the dangers to become more widely known. Thus, none of these three answers affect the argument's conclusion, so none are assumptions.

E) would strengthen the argument, if true, but is not required for the argument's reasons to lead on to its conclusion. Therefore, E) is not an assumption.

Question 6: E
The £50 goes to who has lost the most weight after 4 weeks. Thus we need to calculate the difference in weight at the start, and after 4 weeks for each person. Weeks 1, 2 and 3 are irrelevant.
- Alma started at 71kg, and weighed 64kg after 4 weeks, so has lost 7 kg.
- Brigit started at 67kg and weighed 62kg after 4 weeks, so has lost 5kg.
- Cara started at 63.5kg and weighed 57.5kg after 4 weeks, so has lost 6kg
- Danni started at 66.5kg and weighed 60kg after 4 weeks, so has lost 6.5kg
- Evelyn started at 69kg and weighed 61kg after 4 weeks, so has lost 8kg.

Thus, Evelyn has lost the most weight after 4 weeks, and thus Evelyn get the £50.

Question 7: E
First, we need to calculate the total length of the reign of each King, from the dates given in the table:
- Henry II reigned for 35 Years
- Richard I reigned for 13 years
- John reigned for 17 years
- Henry III reigned for 56 years
- Edward I reigned for 35 years
- Edward II reigned for 20 years
- Edward III reigned for 50 years
- Richard II reigned for 22 year

We can see that all the possible answers have the same groupings when grouping kings according to length of reign. Therefore, we can take these groupings and see how many kings fit into each category. We see that:

- 1 King reigned between 0-15 years (Richard I)
- 3 Kings reigned between 16-30 years (John, Edward II and Richard II)
- 2 Kings reigned between 31-45 years (Henry II and Edward I)
- 2 Kings reigned for longer than 45 years (Henry III and Edward III)
- We see that only Graph E) shows this distribution.

Question 8: D
To answer this question we should break apart each section of the journey described in the question, and see what we would expect the graph to do at this point:
- First, the driver drives *away* from home at a *constant speed*. Therefore, we should expect a straight line (because of the constant speed), heading up and to the right (because both time and distance from home are *increasing*)

- Next, the driver *increases* his speed to a *new constant speed*. Thus, we expect the next section of graph to have a line with a steeper gradient than the last (because the new speed is higher than the last). The line should still be going upwards because the driver is going *away* from home.
- Next, the driver meets a hill, which slows him down to a new constant speed. Thus, we expect the next section of the graph to have a line with a *decreased* gradient (due to the lower constant speed). The line should still be straight, and going upwards and to the right, because the time and distance from home are both increasing.
- Next, the driver stops for several minutes. At this point we expect a *flat* line because there is no movement either away from or towards home. The line continues to the right because time continues to progress whilst the driver is stopped.
- Finally, driver heads back down his route. Thus, we expect the next section of graph to have a line going *down* and to the right, because the driver is now becoming *closer* to home as time progresses.

Only graph D) follows the expected pattern.

- Graph A) does not have a line heading downwards after the flat section, so does not show the driver's return hum.
- Graph B) shows a decrease in speed rather than an increase in speed on the 2nd section of the graph, so does not show the driver's increase in speed.
- Graph C) starts with a flat line, so does not show the driver's initial constant speed.
- Graph E) does not have a flat section, so does not show the several minutes of the driver being stopped.

Question 9: E
C) actually weakens the argument by suggesting that work can be better done with children sitting in groups at tables. B) also weakens the argument by suggesting there is less importance whether children sit in groups or rows, as children may be distracted either way. A) is an irrelevant statement. D) is also irrelevant as the fact that traditionalists favour this arrangement does not affect the conclusion that it is better to do so.
E) strengthens the argument, as if children can see visual aids more clearly this strengthens the idea that they can concentrate more on their work.

SECTION ONE   2011

Question 10: C
The argument describes an increase in police officer numbers, and then concludes that this is why crime is down. C) correctly points out the flaw in this reasoning – crime may have fallen anyway, and may thus not be down to this factor.

B) is not a valid flaw because the question states that crime is down. Thus we must accept this as true for the purposes of a critical thinking assessment, and thus different ways of measuring crime are irrelevant.

A), D) and E) are all irrelevant statements that do not affect the argument's conclusion (D) is incorrect because the question refers to crime *last week*, so any future rises/falls are irrelevant).

Question 11: C
E) is a completely irrelevant statement, whilst B) is also irrelevant as the argument does not say anything about whether risk-taking humans are atypical. It simply mentions variation.

A) is simply a stated reason in the passage, and thus is not a conclusion. D) is also a stated fact in the passage. It can be seen as a conclusion, but it is an intermediate conclusion, which goes on to support the main conclusion from the passage. It is not the main conclusion in itself.

C) readily follows on from the reasoning given in the passage, as a whole. If the reasoning in the passage as true, we have good reason to believe the statement in C). Thus, C) is the main conclusion of the passage.

Question 12: A
To solve this question, we need to work out the prices for a kilogram of coffee from each supermarket:
- ➤ Kostless normally charges £2.50. With the offer of 80p off, this will now be £1.70. 5 X 200 jars will make 1 kg of coffee. Thus, the price for 1kg of coffee is 5 X £1.70, which is £8.50
- ➤ Savemore usually charges £2.50 for 200g. With the offer, we now get an extra 25% extra for the same price. Thus, for £2.50 we will now get 250g of coffee. Thus, 1kg of coffee will cost 4X£2.50, which is £10.00

Therefore, a kg of coffee costs £8.50 from Kostless, and £10.00 from Savemore. Thus, a kg of coffee is £1.50 cheaper at Kostless.

Thus the answer is A)

Question 13: A

In answering this question, we can safely ignore the actual area of each country, and the population density of each country. The % of EU area accounted for by each country is sufficient information, meaning we only need to consider the middle column of the table. This simplifies the question.

To get the minimum number of countries required to account for 50% of total area, we should simply add up countries in size order, beginning with the largest, until we reach 50% of total area:

- The largest country in the table is France, accounting for 14.6% of total area.
- The next largest country is Spain, accounting for 11.4% of total area (we have now accounted for 26% of total area)
- The next largest country is Sweden, accounting for 10.2% of total area (we have now accounted for 36.2% of total area)
- The next largest country is Germany, accounting for 8.1% of total area (we have now accounted for 44.3% of total area)
- The next largest country is Finland, accounting for 7.6% of total area (we have now accounted for 51.9% of total area)

Thus, these 5 countries together account for over 50% of the total area of the EU. Thus, the answer is A).

Question 14: C

This question is simpler than it appears at first glance. Let the books be numbered 1-7, starting on the left (i.e. the left-most book will be Book 1, the second left-most book will be Book 2 and so forth, such that the right-most book will be Book 7).

If we look at Books 1, 2 and 3 under this system, we find that Book 1 is the shortest, Book 2 is taller than Book 1, and Book 3 is the tallest of these 3 books. Thus, we know that any possible view from the left-most side (the direction of the arrow) must show a short book at the front, with a slightly taller book *immediately behind it*, and a taller yet book *immediately behind* this book.

We can see that in the view shown in C), there is a short book, followed by a much taller book immediately behind, and then a book which is between those 2 in height behind this one. Thus, these books cannot be Books 1, 2 and 3 as we observed in the front-view. Thus C) is not a possible view of the books, and C) is therefore the answer.

Question 15: B

A) and C) are both incorrect as the argument states that the universe resembles a vegetable *just as closely* as it resembles a house. It does not say it resembles a statement more, or that it does not resemble a house.

D) and E) are also incorrect as the argument makes no claim as to which is the most likely origin of the universe, and it does not state that any possibility is definitely incorrect. It simply states that we cannot conclude that it was the result of intelligent design.

B) correctly points out the argument's conclusion. The argument is explaining how this particular conclusion of the universe's origin is not justified.

Question 16: A

The argument claims that fanciful claims of aliens causing crop circles can be dismissed on the basis that a more mundane, ordinary explanation exists. A) correctly points out that this assumes that where 2 explanations exist, we should accept the more ordinary one. The argument provides no reasons why we should accept the more ordinary one. Thus, if the statement in A) is incorrect, the argument's conclusion is no longer valid.

B), C) and D) all provide further reasons to believe that aliens have not produced crop circles, and why they are simply a form of rural graffiti. Thus they all strengthen the argument, but are not essential for its conclusion to be valid, and so are not assumptions.

E) is a completely irrelevant statement.

Question 17: C

A) and B) actually strengthen the argument's conclusion, suggesting that houses with burglar alarms are less at risk from burglary because of this, whilst houses/people without alarms are at higher risk.

D) is a completely irrelevant statement, whilst E) is not relevant to the argument's conclusion as it does not say anything about how a visible burglar alarm box affects the risks of a house being burgled.

C) implies that burglars will be more likely to burgle houses with visible burglar alarm boxes, due to the implied value of items in the house. Thus, C) weakens the argument's conclusion. Thus, the answer is C).

SECTION ONE   2011

Question 18: D

First we need to assess what possible digits the month and the date can have.
- The Month can be any number from 01 to 12. Thus, the Month *must* contain either a "0" or a "1".
- The Day can be any number from 01 to 31. Thus, the day *must* contain either a "0", a "1", a "2" or a "3".
- All the possible years have a "2" in, so the day *cannot* include a "2". Thus, it *must* either contain a "0", a "1" or a "3".
- We can see that if the day contains a "3", it must be either "30" or "31". This means that the month must either *not* contain a "1", or not contain a "0".

Now we can assess the years, and see if the numbers taken up leave a possible combination of day and month, without repeating any digits.

- If 2013 is the year, then "2", "0", "1" and "3" are already taken up. This means this year is not a possible answer, as the month has to contain either "0" or "1".
- If 2134 is the Year, then "2", "1", "3" and "4" are already taken up. This means this year is not a possible answer. In order not to share a digit, both the Month and the Day would have to contain a "0", which means they would share a digit with each other.
- If 2145 is the year, then "2", "1", "4" and "5" are already taken up. Thus, the Month will have to contain a "0", as "1" has been used. This means the day cannot contain "0", "1" or "2", so it must contain a "3". This means the day has to be either "30" or "31", both of which already include a digit already used in the date. Thus this year is not a possible answer.
- If the year is 2345, then "2", "3", "4" and "5" have already been taken up. Thus, the Month can contain either a "0" or a "1", as neither of these have been taken up. The day can also contain either a "0" or a "1". Thus, the Month can be any of (06, 07, 08, 09) without sharing any digits with the year or using up both a 0 and a 1. Once one of these is selected, the day can then be selected from of (16, 17, 18 and 19) without sharing any digits with the Year or Month. Thus, 2345 is a possible answer.

Thus, we can see that 2345 is the *next* possible year in which a date will occur which can be written using 8 different digits. This means that D) is the answer. The Year given in E) is now irrelevant, as we have found a year *earlier* than it that is possible, so it cannot be the *next* year in which such a date will occur.

Question 19: B

A) is the view looking at the logo from the right-hand side. We can see the 3 points of the "E" in the Logo, which are on the left, middle and right in this view.

C) is the view looking at the logo from the bottom. We can see the bottom point of the "T" in the centre, with gaps either side, as expected. On the left and right are the wide bases of the "J" and "E".

D) is the view from the top. On the right-hand side we see the upward flick from the "J", with the gap next to it, as expected. The rest of this view comprises the wide top of the sculpture.

E) is the view from the left-hand side. We can see the side of the "J", with the upward flick on the right-hand side of the image.

No viewpoint would give the view seen in B). Therefore, the answer is B.

Question 20: B

This question appears to require some complex calculations of compound interest, but on closer inspection we can see that all the information we require is provided in the table.

Raul's savings are rising at a rate of 6% per year. Thus we can calculate how much they will be worth in years by looking at how much an item will cost in 8 years, and treating Raul's savings as rising in exactly the same manner as 6% inflation. We can see that in 8 years, what was worth $1000 will be worth $1594. Thus, Raul's $10,000 of savings will be worth $15940

We now calculate the cost of the car in 8 years, based on an 8% inflation rate. Raul will buy an equivalent model, so at present it would cost $10,000. In 8 years, with an 8% inflation rate, we can see that an item currently costing $1000 will cost $1851. Thus, an item currently costing $10,000 will cost $18510. Thus, the car will cost $18510.

Thus Raul's savings will cover $15940, of a total expense of $18510.

18510 − 15940 = 2570. Thus, Raul will need an extra $2570 to purchase the next car.

Question 21: A

B) is incorrect as the argument does not assume that police do not protect criminals or protect the public. It simply argues for a "proper focus" to be put on this. This does require us to accept there is not currently a proper focus on this, but it does *not* require us to accept that this job is not currently carried out. Thus, B) is not an assumption.

C) and E) are both irrelevant to the argument's conclusion and are not flaws. D) is incorrect as the argument states that a proper focus could be put on catching criminals *and* protecting the public. It does not claim that catching criminals equates to protecting the public, these jobs could be carried out independently.

A) is a flaw, because the argument is implying that focus on road offences currently detracts from catching criminals and protecting the public. If road incidents are criminal and dangerous, then this conclusion is no longer valid,

Question 22: B

The argument's reasoning can be summarised as "If A happens, B will happen. A hasn't happened, so neither has B. If we want B to happen, A must happen". In the question, "A" is practicing an instrument, and "B" is becoming better at playing it. Only answer B) follows this reasoning, with "A" being working hard at trampolining, and "B" being getting better at it.

In A), "A" does not happen, but "B" still happens, so this is different from the question. In C), "A" does happen, so "B" does happen.

D) follows a different style of reasoning, claiming "If A happens, B will happen. Therefore all who have achieved B must have done A". This is not the same reasoning as the question. E) also follows a different style of reasoning, claiming that "A causes B. A has happened, so we expect B to happen". Again, this is not the same as the question's reasoning.

Question 23: D

The question refers to a situation of people not being penalised for things over which they have no control. D) illustrates this payment, saying Jamuna's payment was delayed by something she had no control over (postal strikes), so she should not be penalised.

A) is claiming that Kelly did have control over her accounts, even though her mistake was not deliberate, so she should be fined. C) is saying that Phillip did have control over his losses so he should not be helped. Neither of these answers follows the principle outlined in the passage.

E), meanwhile, is referring to those responsible for a problem (the financial advisors) being punished (by losing their licenses). B), meanwhile is discussing a situation where those responsible for helping people in a give situation should be left to do so, and others should not intervene. These are both vastly removed from the principle described in the passage.

Question 24: C

To answer this question we must calculate how long each section would take by each road:
- Each part of the journey (Bowbridge to Clatterton and Clatterton to Downland) is 35 miles long. Since the average speed on the motorway is 60mph, each section will take 35 minutes if travelling by motorway.
- From Bowbridge to Clatterton is 30 miles on Buttercup Way. Since we are travelling at 40mph (average), this will take 45 minutes.
- From Clatterton to Downland is 20 miles on Buttercup Way. Since we are travelling at 40mph (average) this will take 30 minutes.

Thus, the fastest route is to travel by Motorway from Bowbridge to Clatterton, and by Buttercup way from Clatterton to Downland. This will take (35+30) minutes, which is 65 minutes.

The slowest route is to travel by Buttercup way from Bowbridge to Clatterton, and then by Motorway from Clatterton to Downland. This will take (45+35) minutes, which is 80 minutes.

Thus, the fastest route will take 15 minutes less than the slowest route.

Thus, the answer is C)

SECTION ONE — 2011

Question 25: B
The secret to this answer lies in examining the sides and angles of the parallelogram in B). In squares, all adjacent sides lie at 90° to each other. Thus, the slanted sides in B) and the horizontal sides cannot be from the same square, as the angles at which they meet are not 90°.

The other way of forming this shape would be to have one square making the horizontal lines, and another on top of it, rotated slightly, to give the slanted lines. As we can see, this would not produce a parallelogram of this shape, because in this instance the width of the square would have to be smaller than its length. This means it would not be a square, whose width and length are equal by definition.
The only way shape B) can be formed is with a square and a *rectangle*. It cannot be formed with 2 squares.

Question 26: D
This question asks us to calculate the change in the *percentage* of households with fewer than 3 children. In calculating this, the total number of households and the average are not necessary. The % figures give us enough information. This simplifies the question as we need only consider a small section of the table.
First we calculate the total % of households with < 3 people in 1971:
- 18% of households had 1 person in 1971
- 32% of households had 2 people in 1971
- Thus, a total of 50% of households had fewer than 3 people in 1971

Next, we calculate the total % of households with < 3 people in 1991:
- 27% of households had 1 person in 1991
- 34% of households had 2 people in 1991
- Thus a total of 61% of households had fewer than 3 people in 1971

The difference between 50 and 61 is 11, so we can see that the change in percentage is 11%. Thus the answer is D)

Question 27: E
E) correctly identifies the main conclusion to the passage.
B) is not a conclusion, and simply a fact stated in the passage, which the arguer then questions and refutes.
All the other possible answers are reasons which go on to support the main conclusion. If we accept answers A), C) and D) as true, they give us good cause to believe that we should see the fall in house prices as a good thing. Thus, E) is the main conclusion.

SECTION ONE   2011

Question 28: C
The argument discusses how working class areas of London drank tea regularly during the first half of the 20th century, and enjoyed a longer life expectancy than other areas.

E) is an irrelevant statement, which has no relevance to the reasoning of the argument. D) is also irrelevant, the argument says nothing about whether tea replaced other parts of a normal working class diet.

A) and B) both assume too much. From the reasons given in the paragraph, we cannot conclude that Tea *definitely* increases life expectancy, as correlation does not necessarily mean causation. Therefore A) is incurred. Similarly, we cannot conclude that everyone should drink tea, so B) is incorrect.

However, from the reasons in the paragraph, we do have cause to believe that tea *may* be beneficial to health. Thus, the conclusion in C) is valid, and C) correctly identifies the main conclusion of this passage.

Question 29: C
The argument discusses how different areas of the country spend different amounts per head of the population over 65 on a certain treatment for Alzheimer's, and then concludes that this means that different areas of the country will have differing levels of treatment for Alzheimer's. At no point is it stated that different areas of the country experience the same incidence of Alzheimer's per head of the population over 65. If this does differ in different areas, then it may well be that different areas spend the same amount *per case of Alzheimer's* on the treatment, making the argument's conclusion invalid. Thus, C) correctly identifies an assumption in the argument.

A) is incorrect as the argument states that the pills are the most effective treatment available. Thus, how they are used is a significant factor in how effective treatment can be. Whether or not they are significantly effective is irrelevant, and thus A) is not an assumption. B) is a completely irrelevant point, which does not affect the argument's conclusion.

D) is irrelevant because the argument refers to the expenditure on the treatment *per head* of the population over-65, so has already taken into account regional variations in numbers of people over 65. E) is not an assumption because it is not needed to be true for the argument's conclusion to be valid.

SECTION ONE   2011

Question 30: A

To answer this question we must work out what we can tell from the table, and then deduce other results as we move along.

Immediately, we can tell several things:
- Rovers have won all their games, and United and County have both only lost one game each. Thus, both teams have lost to United.
- United have won 2 games, and have only scored two goals. Thus, both those games must have been won 1-0 by United.
- Similarly, United have lost one game, and have only conceded 1 goal. Thus, they must have lost that game 1-0.
- This accounts for all of United's scored and conceded goals. Thus, we know United's other game must have been a 0-0 draw.

Thus, at this point we know the following:
- Neither team can have lost the game, as each team has only lost 1 game each, both against Rovers.
- The score must be either 1-0 or 0-0, because these are the only scores which have occurred in any of United's games.

The only possibility is thus that the game between United and County ended 0-0.

Question 31: C

An important point to remember is that each player can play an X **or** an O for their go, thus deviating from the normal rules for Naughts and Crosses.

Thus we need to assess each board and see whether a victory can be guaranteed next turn:
- In A), we can see that if the next player puts a Cross in the bottom-right space, there is no possible victory on the next turn. Thus A) is incorrect
- In situation B) we can see that unless the next player puts a cross in the Top-left, Bottom-left or Top right spaces, no victory can be achieved next turn. Thus, B) is incorrect.
- In Situation C) we can see that wherever an X or an O is placed by the next player, a victory can be achieved on the next turn. Thus, C) is how the current player can guarantee herself victory next turn.
- In situation D) we can see that the next player can prevent a victory on the next turn by placing an O anywhere on the board, thus D) is incorrect.
- In situation E), we can see that if the next player places an O in the left-most two columns, or places a cross in the bottom-right corner, no victory can be achieved next turn. Thus E) is incorrect.

Thus, the answer is C)

Question 32: B

We are trying to calculate how many more people have quit smoking thanks to the nicotine patches than would be expected if they had just been given placebo.

We can see from the table that 24 out of 120 smokers given placebo have quit successfully.

We are told that the researchers predict the *same proportion* of people would have quit without using Nicotine patches as did whilst using the placebo. Since there are 120 people given placebo, and 120 people given nicotine, this means that we would expect 24 people to quit successfully without using the nicotine patches.

In fact, we can see from the table that 56 smokers have successfully quit whilst using the nicotine patches. 56–24=32. Thus, 32 more smokers have quit using nicotine patches than would be expected without the nicotine patches.

Question 33: C

A) and B) are irrelevant. The issue of drivers losing control of the vehicle is irrelevant to which turns they will take whilst what happens in TV car chases may not necessarily be what happens in real-life car chases. Thus, both A) and B) are incorrect.

D) is also irrelevant as further assistance is irrelevant to what direction the drivers being pursued will go.

E) appears relevant, but does not affect how the argument's reasoning leads on to its conclusion. The argument describes how pursued drivers will frequently wish to turn, and how left turns are easier and quicker, and then describes how this means police taking left turns will soon catch up to the criminals. The idea of criminals turning left because that leads to their destination is a separate point from this reasoning.

C) strengthens the argument. C) relates to the idea that even if police follow the path of a criminal, if they wish to catch up they must surely travel at a faster speed. Thus, if C) is true, the argument's conclusion is much more solid.
Therefore, the answer is C)

SECTION ONE — 2011

Question 34: C
The argument describes how the money spent on a new entrance hall is a waste, and how the money should instead be spent on better conditions for scholars who already come to study. However, the argument also states that scholars currently using the museum will continue to do so regardless of how artistic the displays are, and that all scholars who wish to use the museum already do so. This implies that the number of scholars using the museum will not really change much regardless of how the money is spent, so surely spending the money on better conditions for scholars is also a waste? C) correctly points out this flaw in the argument's reasoning.
A) and B) actually strengthen the argument, adding reasons that spending money on the entrance hall may not produce significant results, so contributing to the argument that this expenditure is wasteful. E) also collaborates with the argument, arguing that the money should be spent on current visitors, not attracting new ones.
D) is an irrelevant statement, not relevant to the argument's conclusion.

Question 35: A
The argument discusses the effects of population growth from unrestricted immigration policy, and then concludes that this will cause Extremist political parties to gain political ground. Answer A) correctly identifies this main conclusion.
E) is irrelevant to the argument's conclusion. The fact that opposition parties believe immigration is responsible for population growth does not affect how extremist parties will benefit from the rise in population.
B) and C) are a mixture of facts and opinions stated in the passage, whilst D) is an intermediate conclusion that goes on to support the main conclusion in A).

Question 36: E
Each Pizza has a fixed charge. Let the fixed charge be termed X.
Each pizza also has a variable charge, proportional to the amount of ingredients used. Let the variable charge for a small pizza be termed Y.
- A small Pizza will have 1Y as its variable charge.
- A regular pizza will have 2Y as its variable charge, as it uses twice the ingredients of a small pizza.
- A large pizza will have 4Y as its variable charge, as it uses twice the ingredients of a regular pizza.

Thus we are looking for a set of prices which can be described as (X+Y), (X+2Y) and (X+4Y), where X and Y are constants. Only the prices in E) follow this pattern, with X and Y both being £1.

SECTION ONE 2011

Question 37: A

Without the extra item, the total price would have been £74.60. We can see from the question that for an order of this price, there would be a £12.00 charge for postage and packing (for orders between £50.00 and £74.99). Thus the total price would be £86.60.

With the extra item, the total price for the order is now £75.85. We can see from the question that the normal charge for postage and packing for such an order would be £15.00. However, since we now have 8 items, the postage and packing charge is halved. Thus, a £7.50 charge is levied for postage and packing. Thus the total price is £83.35

86.60-83.35=3.25. Thus, the extra item took £3.25 off the total price.
Hence the answer is A).

Question 38: A

Let the number of brochures be X. We can now describe the price of each structure as follows:
- Price structure 1) cost simply equals X (Y), where Y is the cost per brochure.
- Price structure 2) cost equals W + X(Z), where W is the fixed price and Z is the cost per brochure. We know that Z is smaller than Y.

Thus we can see that:
- Price structure 2 will start out as the most expensive, due to W.
- However, because Y is larger than Z, as the number of brochures increases, the price of Price structure 1) will increase at a larger rate than the price of price structure 2).
- Thus, as the number of brochures increases, Price structure 2) will be comparatively less expensive when compared to Price Structure 1), until we reach a point where they are the same (let this number be termed Equiv).
- Above point Equiv, Price Structure 2 will be cheaper.

Thus, knowing point Equiv (the number of brochures for which each price structure is the same) is sufficient to know which price structure will be cheaper. If Point Equiv is greater than 500, price structure 1) will be cheaper. If Point Equiv is less than 500, Price Structure 2) will be cheaper.

Thus A) is the answer.

SECTION ONE — 2011

Question 39: E

The passage describes how both mobile phones and masts emit radiation, and describes how mobile phones emit much more. The passage then discusses how these levels are both below what the guidelines say is dangerous, but that we cannot be sure if these guidelines are accurate. From this, we cannot conclude that mobile phones put people's health at risk, we can only conclude that we can't be certain they don't. Thus, A) is incorrect. Neither can we conclude that the guidelines are incorrect, or those living close to a mast over-estimate the risk. Thus, B) and D) are incorrect.

C) is mistaken. The passage has said that there is less radiation emitted from a mobile phone than a mast, and described how much lower the radiation level is.

However, occasional/infrequent mobile phone use may result in less radiation than living near a mast. Thus, C) is not a valid conclusion.

However, E) identifies a valid conclusion. Since the radiation levels are lower from a mast, if *extensive* mobile phone use poses no threat, it is *unlikely* that living near a mast does.

Question 40: D

The passage describes 2 purposes of University education, and then mentions there are others. It does not claim that any of these is the primary purpose, so A) and B) are incorrect.

C) and E) are not relevant to the argument's conclusion (which is that University should not simply consist of forcing through as many graduates as possible at minimum cost), thus neither are assumptions.

D) correctly identifies an assumption. The argument discusses how we should not simply force through graduates at minimum cost. However, it is not stated that cost-cutting policies will stop Universities achieving their overall purpose. If this is not true, the argument's conclusion is no longer valid. Thus, D) is a valid assumption in the passage.

SECTION ONE   2011

Question 41: B
The passage is giving reasons that the maintenance of a First Class section on trains is detrimental to the service enjoyed by *most* passengers (those travelling in standard class). We can quickly see that if B) is true, then the opposite is actually true, and First Class is of benefit to the rest of the rail service, thanks to the revenue it produces. Thus, B) weakens the argument.

A) and E) do not affect how the maintenance of a First Class service affects the facilities in Standard class, so neither of these affect the argument's conclusion.
D) is an irrelevant statement as the passage has said that First Class is not indefensible because of the expense.
C) actually strengthens the argument, as it reinforces the idea that the majority of people are travelling in standard class, thus contributing to the conclusion that *most* passengers suffer from the First Class provision.

Question 42: E
If the piece of card is folded up, then we can see that the triangle extension on the bottom-right of the net will align with the top section (with the diagonal line separating shaded and non-shaded), such that the tip of the non-shaded section of the triangle is touching the tip of the *shaded* section of the top section. Thus, A) will be a possible view from the right-hand side of the shape once it is made.
We can readily see that B) will be a possible view, because we can see that the triangle extension on the bottom right is already attached to the bottom section (with a horizontal line separating shaded from non-shaded), as seen in B).

C) is also possible, because the bottom-right triangle extension will align with the middle section (with a vertical line separating shaded from non-shaded), such that a non-shaded edge of the triangle is aligned with the shaded section of the middle section, as seen in C).

D) is a possible view because the middle-left triangle extension will align with the top section (with the diagonal line separating shaded from non-shaded), such that a non-shaded edge of the triangle is aligned with the *shaded* side of the top section, as seen in D).
E) is not a possible view. We can see that the bottom-right triangle is aligned with the bottom section in the opposite way to that seen in E) (i.e. the shaded and non-shaded sections are reversed compared to how they appear in E). Meanwhile, the left triangle extension will align with this section such that a non-shaded edge aligns with the bottom section, which is not what is seen in E). Thus, E) is not a possible view.

SECTION ONE 2011

Question 43: C

First we must calculate the number of students studying Spanish. We can see that Spanish features twice in this table, in the Second row from the top, and in the right-hand column.

Thus, we simply add up the numbers in the second row from the top, and in the right-hand column. This gives us a total of 40 (14+9+13+4=40).
Now we see how many students are studying Spanish *and* French. We see from the left-hand column of the table that there are 14 students studying Spanish and French.

Thus we simply calculate 14 as a percentage of 40: $\left(\frac{14}{40}\right) X 100 = 35$
Thus 35% of Students studying Spanish are also studying French.
Hence, the answer is C)

Question 44: D

We can see that the different lines readily produce 5 triangles, immediately apparent. Thus, this accounts for 5 possible routes. The other routes are as follows:
- Start at the central intersection, proceed to the top left corner, then proceed to the bottom-most corner, then proceed back to the intersection. A mirror image route is also possible using the top-right corner. Thus, this accounts for 2 more possible routes.
- Start at the central intersection, proceed to the Top-left corner, then proceed to the bottom-most corner. At this point we then proceed in the direction of the top-right corner, but stop where there is an intersection with a path leading to the central intersection. We then head back to the starting point, along the path to the central intersection. Again, an opposite route is possible, beginning by heading to the top-right corner. Thus this accounts for 2 more possible routes.
- Start at the central intersection, and proceed to the top-left corner, then head to the top-right corner. At this point, we proceed in the direction of the bottom-most corner, but stop at the same intersection described in the last possible route, and head back to the central intersection. Again, there is an opposite possible route, beginning be heading to the top-right corner, then to the top-left corner. Thus, this accounts for two more possible routes.

Thus, we have accounted for 11 possible routes.
Thus, D) is the answer.

Question 45: D

The argument discusses how study leave, whilst benefitting hardworking students, does not produce the best results. It then finishes by saying "it is time to ditch study leave". This last sentence is crucial. Without this last sentence, C) would probably be the main conclusion, as there would be no suggestion in the argument that study leave should be abandoned, simply that it is not the best way to attain high results. However, the last sentence means that the argument *does* conclude that study leave should be abandoned. Thus, D) is the main conclusion. This means that C) is now an intermediate conclusion, going on to support the conclusion given in D).

A) is a reason given to support C), and thus contributing towards supporting D). Thus A) is not a conclusion in itself. E), meanwhile, adds further strength to the argument but is not a direct conclusion from it, as the argument has not claimed that traditional methods need to be updated.

B) actually weakens the argument by picking up on the fact that some students do very well from study leave, thanks to their own hard work and organisation.

Question 46: E

The argument is discussing how these lands were not obtained through just means, and so should be redistributed. The claim that they should be confiscated *directly leads on* from the fact that they have not been voluntarily redistributed, *because* of this fact.

Thus B) is incorrect, as B) implies that a different set of reasons is behind why they should be confiscated. D) is also incorrect, as the argument does not make any *legal* claims about why they should be confiscated, the argument is talking about social justice, and therefore referring to moral reasons, not legal reasons. A) is also irrelevant as the argument gives no economic reasons why they should be redistributed.

C) and E) both pose the issue of confiscation as leading on from the fact that the lands have not been voluntarily redistributed. However, only E) states that the confiscation should happen *because* they have not (*since* they have not). Thus, E) logically follows completes the argument.

SECTION ONE — 2011

Question 47: A

The argument's reasoning can be summarised as "All things that are A have/do/are B. This item is A, so must have B". Here, "A" is being a tree, and "B" is having leaves.

A) most closely follows this reasoning. Here "A" is all leaves growing, and "B" is those leaves being on trees. Thus, we conclude that where there are leaves, they must be on a tree. Thus, if a plant has leaves, it must be a tree.

B) reasons that "All of A has B", and the incorrectly concludes that if something has B, it must be A. This is not valid, as there may be other things apart from "A" that also have "B". D) also follows this incorrect reasoning, but with "oak" replaced by "tree" and "tree" replaced by "plant".

C) reasons that "All of A has B", then observes "This instance of B is C", and then incorrectly concludes that "All of A must be C". This is completely invalid, as there is no suggestion that everything that only C gives rise to B.

E) is also incorrect, claiming that "All of A has B, and that this example is not A", before incorrectly concluding that it does not have B. This follows the same flaw as B) and D). There may be other things apart from A that have B.

Question 48: B

First we calculate how much the charge would be for the detailed information on the 5 companies:
- Each request would be charged at 50c
- Thus, a total of $2.50 would be charged for detailed information on 5 companies

Then, Megan has used the other services 42 times. Each other service costs 10c, so it is irrelevant which services she has used. Thus the total charge for all other services would be $4.20

The question states that Probe's charges do not include the cost of phone calls, and then asks how much Megan will be billed by *probe*. Thus, the call charge need not be considered, as this will not be included in her Probe bill. The phone company will bill her for the call charges.

Thus, Megan will be charged $2.50 + $4.20, which is $6.70.
Thus, the answer is B)

Question 49: C
We are asked how many different possible combinations of course there are for a student taking Drainage. Since there is now one of the options filled, we can quickly work how many combinations would be possible without the restrictions given in the question:

➤ There are 6 other courses. Thus, multiplying 6X5 gives us the total number of ways in which each option could be listed with each of the other options. This gives us 30 possibilities.

➤ When we are thinking of possible course combinations, this number needs to be halved, as it will count each possible course twice (e.g. this number will include "Bricklaying and Damp-proofing" AND "Damp-proofing and Bricklaying" as possible course combinations. We can quickly see that both of these would result in the same combination of courses).

➤ Thus, without the restrictions given in the passage, there would be 15 possible combinations.

Now we can work out how many possible course combinations are made not possible by the restrictions given in the passage:

➤ Drainage and Damp-proofing may not be taken together with Bricklaying or Plastering. This removes 2 possible course combinations (Drainage/Damp-proofing/Bricklaying and Drainage/Damp-proofing/Plastering)

➤ If Joinery is taken, Flooring must also be taken. This removes 4 possible course combinations (The combinations which now cannot be taken are Drainage/Joinery/Bricklaying, Drainage/Joinery/Damp-proofing, Drainage/Joinery/Plastering and Drainage/Joinery/Roofing)

Thus, 6 possible combinations are removed by the restrictions given in the question. Thus the number of possible combinations is 15-6, which is 9.
Hence, the answer is C)

SECTION ONE — 2011

Question 50: C

Let the fixed postal charge be termed "Y".

There is a £2.00 discount per book versus shop prices. Thus, the total savings in £ versus shop prices can be described as: 2(Number of Books)-Y.

We can see from this equation that as the number of books increases, the savings will increase *in multiples of 2*, as each book adds another £2.00 saving (and we cannot add a fraction of a book to the order, we can only add whole numbers of books). Thus, graph B) cannot be correct, as in this graph the saving increases by more than £2 for each book that is added.

We can also see that thanks to the fixed postal charge, the saving for 1 book **cannot** be higher than £2, as there will be a £2 saving *minus* the postal charge. Graphs A) and E) both show the saving for 1 book being £2 or higher, thus both of these graphs cannot be correct. We can also see that in Graph A) the saving increases in increments of less than £2 per book, adding further reasons why this cannot be correct.

Additionally, we can readily see that answer D) is incorrect, because this shows the saving being higher for 1 book than it is for 2, 3 or 4 books. This cannot be possible, as we can see from our equation that the saving will increase as the number of books does.

Graph C) is the only graph which fulfils all these criteria, with a saving of less than £2 for 1 book, and the saving increasing at a rate of £2 per additional book.

END OF SECTION

Section 2

Is the general understanding of science damaged by the way it is presented in the media?

When answering this question you should be careful to define your terms. Specify which forms of media you will be discussing (these should be the ones you think most relevant to the question) and what you consider to be 'damage' to general understanding – is any kind of misunderstanding damaging, or only that which causes actual harm to an individual or the general public? Be wary of generalising 'media' as a whole also; consider whether some forms of media have been more damaging than others, or whether it's a case of good science and bad science, both of which can occurs in any form of media.

Introduction:
- You should use your introduction to give an overview of how you believe the media presents science, and what media you will be referring to. Since the 'general understanding' of science through media is mostly through popular media such as pop science books and documentaries, you will most likely want to start there and give a few examples of the kind of media you will be discussing.
- You may wish to describe the way science is presented in media as any number of: accurate, inaccurate, sensationalised, biased, unbiased, varied.
- Once you have established what you here mean by 'media' and how you think it presents science, clearly state your opinion on whether the general understanding of science is damaged by this, and the arguments you will use to defend this view, as well as the counterarguments to these and your reasons for dismissing them.

Potential arguments for:
- Popular media often simplifies scientific concepts so they can be understood by more people, but as a result the concept is twisted and misunderstood.
- The media is sensationalistic and presents scientific breakthroughs or a discovery of some new piece of evidence as more significant and conclusive than they actually are. Where a researched might say "study Y has found evidence consistent with conclusion X", a newspaper may print that "scientists prove theory X!"
- Some presentation of 'science' in popular media may be misleading or downright inaccurate – think of the arguments against vaccinating children that point to the tiny, harmless mercury content in vaccines, or use fear

mongering and hyperbole. Although we can dismiss this as 'bad science', the extent to which the public has been taught to believe representations of science presented on TV or in books uncritically is damaging, and a result of a lack of emphasis on transparent sourcing of information in these mediums.

Potential arguments against:
- Any moderately accurate representation of science in media accessible to the public (popular science books, documentaries, etc., as opposed to scientific journals) aids a general understanding. Although the concepts may be simplified to make them accessible and understandable, this is still preferable to no understanding at all, and people generally know they do not derive a completely accurate understanding of the subject from these sources alone.
- It is fine for the general public's understanding to be slightly inaccurate, or communicated primarily through, say, analogy rather than a more precise explanation. There are certain things the public needs to have some understanding of – how vaccines work, for example, and why they're important – so they can make an informed decision, and this understanding should be communicated in a way that is accessible to someone without a biology degree.
- The simplification of scientific concepts that occurs in popular media is no worse than the simplification that occurs at various levels of education. For example, until a certain point, school children are taught that light always travels in straight lines, because this is usually true, and is all the information they need to understand things like how we see colour, reflection, and so on. If they pursue physics to higher levels, they will learn that light bends near sources of very strong gravity – but this is not something you need to know if your physics education doesn't go beyond more simplistic concepts.
- It is impossible to say that all media presentation of science is damaging because it is such a vast and varied industry. While there are undoubtedly some inaccurate, overly simplistic, or sensationalist representations of science in the media, there are just as many representations that are highly beneficial to the general public. To say that the media in general is damaging is overly simplistic itself.

Conclusion:
- Summarise what constitutes media, and how it generally presents science.
- If you have given a more nuanced view – that some presentations are damaging while others are not – be sure to make that perfectly clear in the conclusion, and outline what makes the difference between a damaging and beneficial presentation of science.

SECTION TWO — 2011

Do patent laws encourage or hinder development?

This essay is easy to structure well. Begin by clearly presenting the case for your side, then explain the potential counterarguments to your points, and your reasons for dismissing them. It is worthwhile to note that the question assumes patent laws have either a positive or negative effect on development; it is perfectly valid to question this assumption by asserting that they have no impact on development. If you wish to take this route, make this point in your introduction, then present the case for them encouraging development, then the case for them hindering development (or the other way round; the order is unimportant if you take this route), and refute each side. Patents are a controversial topic with many arguments on both side; be careful not to get off topic, and limit your argument to that which concerns development specifically.

Introduction:
- Begin by outlining what you believe 'development' to be. One potential definition is "the process of growing and becoming more mature, advanced, or elaborate". Given that this question applies the concept to patents, the development will be in technology, including electronics, manufacturing processes, recipes, medication, etc. – in short, anything that can be sold.
- It is worth also giving a (very) brief explanation of what a patent is – "a government authority or licence conferring a right or title for a set period, especially the sole right to exclude others from making, using, or selling an invention" – or some other explicit indication that you know what it is.
- Clearly state your position and outline the main arguments you will be using to defend it, as well as counterarguments to these and why you feel they should be dismissed.

Potential arguments for encouraging development:
- Patents give an individual sole possession over their intellectual work. This motivates people not only before they have developed a piece of technology (since they know they will get the recognition for and ownership of the product they deserve), but after they have developed it, they will be motivated to keep working on it, as they are the only ones who can.
- Patents of technology encourages the competition to develop their own. Without patents, everyone would simply use the technology developed by other people, with no motivation to create new, better products.
- Patents allow developers of new technology to ensure they can make a profit, by being the sole producer of a certain product. Without this guarantee, people wouldn't be as motivated to develop new technology.

- Patent holders can license their technology to other developers, so there is no risk of completely limiting development if the patent holder decides to go no further with their technology, or if other developers have the potential to improve it.
    - **Counterargument**: this may be the case, but licenses can be very expensive, and represent a cost that at least some, if not many, developers will not be able to afford, and so naturally limits development – and this is if the patent holder is willing to license their patented technology at all.

Potential arguments for hindering development:
- Patents disallow the possibility of new developers either improving on existing designs, or using small pieces of technology as component parts in new products; patents prevent these products being developed at all.
- Patents allow one company to hold a monopoly over a certain industry. This is not only bad for consumers in everyday industries (as monopoly holders can set prices according to what will maximise their individual profits, rather than at competitive rates), but can be fatal when applied to the healthcare industry. Medication patented and monopolised by one company can be vastly too expensive for some people to afford; consider the millions of poor people who die from HIV every year because they can't afford the patented medication that could save them. This many people dying preventable deaths as a result of patents naturally hinders development.
- The development of new technology rarely happens by developing completely new products. Rather, people tend to modify and improve on existing products, but this is too similar to the old technology to be legal if it is patented, so patents prevent this kind of development.
- People don't develop new technology for the sake of profit alone, they do so because of natural human inventiveness and drive, and a desire to fill a need or want not currently being fulfilled. They would continue to do this even without a guarantee of profit as a result of patents and, indeed, they already do – even with patents, there is no guarantee their efforts to develop new products will be fruitful, or successful in the market, and yet they continue to develop.
- The cost of developing new technology – requiring intellectual labour, manufacturing of various designs, testing, etc. – is very high, but the cost of reproducing technology once we know how to make it is relatively very low. If technology is all patented, the overall cost of production in any given industry is very high, since each producer must develop and produce their own technology. This is particularly concerning in fields such as medication,

for the reasons discussed above, but hinders development in all fields by raising the costs of production.
- These higher costs of production further hinder development by preventing *new*, potentially very good, producers from entering the industry – the costs of entry are too high. If they could use the technology already available to them, they could work on developing only their one or two particular areas of interest, and progress would be made there. As it is, in order to enter an industry developing new technology, they would need to develop new forms of all the technology they need.

Conclusion:
- Restate your position and summarise the main arguments you have used to defend it.
- Be sure to link your conclusion back to the question by referring to the concept of development, and the impact patents have on it.

Do coalitions necessarily adopt policies which unite party leaders but alienate party followers?

The two most important words to note in this question are 'alienate' and 'necessarily'. The concept of alienation is a tricky one, so be sure to give a clear definition of how you will be using the term in the introduction, and refer to a concrete concept of alienation (rather than some other 'bad' outcome for party followers) throughout your essay. 'Necessarily' is important because the question is asking whether this will *always* be the case, whether alienation of party follows *must* follow from coalitions which will *always* adopt policies which unite party leaders. It is possible to answer this question, for instance, by arguing that coalitions *tend* to have this result, and *usually* will, but it is not *necessarily* the case. This essay can be structured simply by presenting your case, considering the counterarguments, and giving your reasons to dismiss them.

Introduction:
- The term to give a clear definition of here is 'alienate' – one potential definition of alienate is "to make someone feel isolated or estranged".
- It is also important to give a clear indication that you know what a coalition is. This need not be a precise definition, and should be implicit throughout your whole essay, but be sure to make sure this is clearly conveyed to the reader.
- Clearly state your viewpoint and outline the arguments you will use to defend it, and your reasons for dismissing counterarguments.

Potential arguments for:
- In a coalition, there usually needs to be some form of agreement between party leaders on any legislation; this means there will naturally have to be compromise on party policy, and the party followers, presumably broadly in support of their party's original policy promises, will naturally feel alienated by the leaders' move away from the party's principles.
- Having voted for their party on the understanding that, if they get into power, they will enact their policies, party members will feel alienated when seeing their party leaders agree to policy not in line with their principles.
- Coalitions will adopt policy that unites party leaders because this is the *only* policy they are able to agree on: where party leaders' are too opposed to changing their established policy to compromise, no policy will be able to be enacted. Where this occurs, in the UK at least, legislation is often left to an open vote in the House of Commons (rather than members of both parties being whipped into voting one way or another); this is not the coalition adopting a policy, but rather legislation emerges from the components of the Commons and their say on proposed Government policy.

SECTION TWO — 2011

Potential arguments against:
- Though we may observe any number of coalitions that have done this, this is not even close to establishing that they *necessarily* do.
- It is not the case that there needs to be some form of agreement between party leaders on legislation. In coalitions where there is a dominant member (i.e.: a coalition in which one party has a large plurality, and the other a minority) and there is a clear leader of the coalition, they are able to dominate policy and adopt ones consistent with their followers' preferences. An example of this is the 2010 UK coalition between the Conservatives and the Liberal Democrats.
- Compromise on policy is not likely to make party members feel alienated. They will understand that their party does not have the power to enact policy without the approval of their coalition partner(s); so long as their party leaders continue to attempt to enact party policy, they may be frustrated and disappointed with the result, but will not feel alienated.
- Coalitions do not necessarily adopt policies that unite party leaders. In Germany, for example, coalition governments are the norm, rather than single party governments as in the UK, and so open disagreement between coalition members is considered usual.
- Further to this, where coalitions are more frequent than they are in the UK, party members are perhaps less likely to feel alienated by leaders compromising on policy, as it is expected even when the electorate votes for their preferred party.
- Parties tend to enter into coalitions with other parties at a similar position on the political spectrum as themselves. Their policies may therefore not be too dissimilar, and compromise on some or many issues may not be severe enough to cause feelings of alienation (which includes isolation and estrangement from) the party among members.

Conclusion:
- Link your conclusion back to the question by referring to the nature of coalitions, and whether they can have a fracturing impact on party members.
- When answering questions of this nature, it can be nice to, at the end of your essay, make a short concluding comment (only a sentence or two) on the debate itself – for example, a comment on how significant alienating party members may be to the party, to democracy, etc.

SECTION TWO — 2011

Should we have a right to choose when and how we die?

This question is highly controversial, and as such provides a great many points to make on both sides of the argument. Be sure to note the interesting concept of a 'right', which implies that anyone should have this freedom, not only those to whom the term 'euthanasia' would apply, as many people would otherwise assume. This essay is easy to structure well: lay out the outline of the argument in your introduction, then make the case for your view point. Follow this with counterarguments, and give your reason for opposing each of them.

Introduction:
- Define the concept of a 'right' – one such definition is "a moral or legal entitlement to have or do something". If someone has the right to choose when and how they die, then it is something they *prima facie* should be allowed to do.
- Clearly state your position and outline the arguments you will use to defend it, as well as the counterarguments to these and your reasons for dismissing them.
- If you are going to argue that people do have a right to choose when and how they die, you may want to use your introduction to quickly outline any limitations on this – for example, if there are limitations on exactly how they may do it (suicide in public spaces, for example, being prohibited). If you like, you may expand on this in greater detail later in the essay, but given the limited time it is also fine to focus on arguments surrounding whether we have any right to choose to die at all.

Potential arguments for:
- Many people who want to make the choice to die are terminally ill and/or in a lot of pain that cannot be prevented. Allowing them to die is the kindest thing to do, so they may do so as painlessly and with as much dignity as possible.
- The choice between a lot of pain for a short amount of time and then death, and immediate death, is not a choice anyone but those in the situation can relate to, so it is not up to anyone but them to weigh in on the decision.
  - **Counterargument**: is this tantamount to a generic 'right to die'? Surely a *'right'* to choose how and when you die must be applicable to anyone, not only the terminally ill or those in a great deal of pain. This argument does not concern a right to choose how and when you die, it concerns a right to euthanasia, which is a related, though different, debate.
- People should have completely bodily autonomy; they know what is best for themselves better than anyone else, and should be able to carry this out.

Indeed, even if they *don't* know what is best for themselves, they ought to be able to make choices about their bodies, which are wholly their own.
- **Counterargument:** what about people who can't make even good decisions for themselves? We do not allow, for instance, small children to decide their own medical care; are there are cases of adults who are in an inferior position to someone else to make medical decisions?

➢ We already allow, and mostly agree with allowing, people to decide in advance when they would like to stop being assisted in living ("do not resuscitate" orders, for instance), and to make the choice to stop receiving treatment if they decide the effort of continuing to get better outweighs the benefits. Allowing people to make the active choice to die is no morally different from this, but does save them the pain of having to die slowly without medication, or wait until their condition is severe enough that simply not being resuscitated will kill them.

Potential arguments against:
➢ Giving people the right to choose when they die is open to abuse by, say, people who no longer want the responsibility of looking after elderly or sick family members, or who want their inheritance; they could bully or manipulate these family members into 'choosing' euthanasia. This is especially relevant given that a disproportionate amount of the people who would choose to die would be very old or sick, and so vulnerable to such manipulation.
- **Counterargument:** isn't this just an argument for restricting how one may enact that choice to die? Perhaps they must sign a waiver and discuss their choice with a doctor or psychiatrist (perhaps one specially trained in such matters) before going through with it, to ensure such coercion does not occur. This is not an actual argument against the right to die.

➢ People who would choose to die are not in their right minds, since no one in their right mind would choose to die; they ought to be protected from themselves and prevented from acting rashly.
- **Counterargument:** this argument is circular and unfounded. True, the choice to die seems farfetched to a lot of people, but that is because most people do not have a desire to die. This desire does not come only from being of unsound mind, but also from being in extreme, unpreventable pain, or facing inevitable and imminent death anyway – both of which are scenarios the average people cannot relate to. To label this as evidence of insanity from which the person needs to be protected is judgemental and condescending.

- **Counterargument:** further, who are we to say that, say, a depressed person is not 'in their right mind' or this is not 'what they really want'? They are still capable of rational thought and making the best decision for themselves; if their mental pain is so severe they decide they would like to die, how is this different from the patient in an extreme amount of physical pain who would like to die?
➢ People should only have the right to die in certain situations – for example, if they are terminally ill and want to avoid the pain of wasting away slowly. Allowing it at any time has too much potential to be exploited by dangerous, manipulative individuals, and for death to occur as a result of neglect – not taking into account someone's severe depression when they 'choose' to die, for instance.
  - **Counterargument:** is it really a 'right' if it's only allowed in some circumstances? No other person has the right to choose when another is in a severe enough situation to warrant control over their own life.
➢ Giving people the right to choose how and when they die has implications for people who are not able to kill themselves; this 'right' will effectively mean they have the right to assisted suicide, i.e.: they have the right to insist that someone else kill them. This is not a right to which anyone is entitled.
➢ Dying doesn't only affect the individual who dies; family, friends, involved medical professionals, all deal with hurt, loss, grief, and guilt, among other emotions; is it right for the dying person to inflict this on them?
  - **Counterargument:** a great many choices we make over our own bodies may cause distress to another person (e.g.: people who have abortions distress those who oppose abortions; transgender people who seek gender confirming surgery distress transphobes; people who get tattoos may upset their parents), this is not a reason to undermine their fundamental right to bodily autonomy.

Conclusion:
➢ If you have taken a nuanced view – for example, if you have argued that people have the right to choose when and how they die only in certain situations, e.g.: in the case of a terminally ill patient – be sure to make that explicit in your conclusion, and outline what constitutes an appropriate situation for this right to be made available.

END OF PAPER

2012

Section 1

Question 1: D
D) correctly identifies the main conclusion of the argument. If we accept the reasons given in the passage as true, they support this conclusion.

A), B) and C) are all reasons given in the passage which support this conclusion. E), meanwhile, is neither a reason nor a conclusion from the passage. The passage makes no reference to whether speaking a foreign language is easier or harder than sign language

Question 2: A
Stewart and Michael live 8 kilometres apart so to meet they will need to travel a total of 8 kilometres in total. It only matters who travels each distance in terms of the time it will take; as long as the total is 8km as they can meet anywhere between the two houses. Michael can cycle 5/3 as quickly as Stewart can run so Michael can travel 5/3 the distance as Stewart in the same time.

Therefore when they meet, Michael should travel 5/3 as far. Hence Michael should travel 5 kilometres and Stewart should travel 3 kilometres. It takes Stewart an hour to run 12 kilometres so it will take him quarter of an hour, or 15 minutes, to run 3 kilometres (we can verify this by checking how long it takes Michael to cycle 20 kilometres). Hence the answer is A.

Question 3: E
E) correctly identifies a flaw in the argument's reasoning. If we accept E) as true, then the argument's conclusion no longer logically follows from its reasoning, and thus E) is a valid flaw from the argument.

D) is not a flaw, as possible methods to solve the issues created by a lack of demand for meat and dairy do not detract from the fact that government advice to eat more fruit has been a cause of this. A), meanwhile, is a matter of opinion, and does not directly affect whether the argument's reasoning lead on to its conclusion. B) and C) are irrelevant to the argument's conclusion.

SECTION ONE 2012

Question 4: D
The argument says nothing about whether the risk outweighs the benefit, and for the purposes of a critical thinking assessment like the TSA, we should accept the reasons given in the argument as true. Therefore B) and C) are incorrect. E) is also incorrect as the argument makes no claims about whether government should take a role in regulating piercings.

A), meanwhile, is incorrect as the fact that specialist piercers account for few hospitalisations does not necessarily lead on to the conclusion that they give good advice about caring for piercings.

D), however, can be reliably concluded, as it does readily follow on that people would be well advised to go to a reputable piercer if these account for few hospitalisations.

Question 5: D
At no point is it stated that women are as likely as men to possess the skills required for senior working positions. However, if this is not true, the argument's reasons do not logically lead on to its conclusion. Therefore, D) is a valid assumption in the passage.

B) and E) do not affect the argument's conclusion and are thus incorrect, whilst C) is not required to be true for the argument's conclusion to be valid, so is not an assumption.

A), meanwhile, actually weakens the argument, if true, so is not an assumption.

Question 6: B
When the daughter is not at home, they spend £120. 70% of this is the food bill, which is £84. 20% is household goods, which is £24. 10% is store cupboard essentials, which is £12.

When the daughter returns, the weekly food bill increases by 20%. 20% of £84 is £16.80, so the food bill is now £100.80. Spending on household goods goes up by 5%. 5% of £24 is £1.20, so the household goods bill is now £25.20. They still spend £12 on store cupboard essentials.

Hence the total food bill is £100.80+£25.20+£12.00 = £138.00. Hence the answer is B.

Question 7: C

On Wednesday afternoon, there must be an English teacher and a Spanish teacher invigilating. All of the English teachers are female, so Marjorie or Jenny must invigilate with a male Spanish teacher. The only male Spanish teacher is Keith, so he must invigilate on Wednesday afternoon.

We also know that each person can only invigilate once a day. In the morning, there is a Maths exam, and Marjorie is the only Maths teacher. Therefore she must have invigilated in the morning, which means she cannot invigilate in the afternoon.

Hence Jenny must invigilate in the afternoon. Hence the invigilators are Keith and Jenny.

Question 8: C

Cutting the cylinder lengthways down the middle allows the views seen in A) and D). Cutting widthways across the middle of the cylinder allows the view seen in B). Cutting diagonally across the length allows the view seen in E). There is no cut which produces 2 equal halves which allows the view seen in C).

Question 9: C

C) would weaken the argument, as it suggests that the risk present to endangered animals is small and not significant, since endangered bear populations have not been affected by the growth in hunting clubs. B) does not necessarily weaken the argument, because the fact that few people take part in the activities does not definitely mean that they will not have an effect on the populations of endangered animals.

D) and E) actually strengthen the argument, suggesting that the clubs are not necessary and that they have a larger effect on wildlife by killing more animals than their targets. A) is an irrelevant statement.

Question 10: E

E) correctly identifies a flaw in the argument, pointing out that even if laptops run out, they will still be of some use to passengers, and therefore the argument they take up a "disproportional" amount of space may be invalid.

C) and D) are completely irrelevant to whether the laptops take up a disproportional amount of space, whilst A) does not affect the conclusion because the argument has stated this should happen "regardless" of whether the laptops are used for personal or business use.

B) meanwhile actually strengthens the argument by providing reasons for laptops being carried in the hold rather than as carry-on items.

SECTION ONE 2012

Question 11: A

A) correctly identifies the main conclusion of the argument. All the reasons provided come together to support the idea that parents should to more to regulate what their children watch on TV.

C) and D) both contradict points of the argument, and so are not conclusions.

B) and E), meanwhile, both serve to reinforce the main conclusion. If parents are not concerned by what their children watch on TV, and some do not believe it is causing their children to behave badly, then this helps support the idea that parents are not doing enough to control what their children watch on TV, and should "do more"

Question 12: B

45 people had neither maths or economics as part of their degree, so 230-45 = 185 people must have had either maths or economics, or both, as part of their degree. 127+89 = 216 so there is an overlap of 216 – 185 = 31 people. Hence 31 people must have had both maths and economics as part of their degree. Hence the answer is B.

Question 13: A

1 playing board is needed for each box, so 8 are needed for a carton. There are 98, so this makes 12 cartons. 50 question cards are needed for each box, so 400 are needed for a carton. There are 5000 question cards, so this makes 12 cartons.

6 playing pieces are needed for each box, so 48 are needed for a carton. There are 1000 playing pieces, so this makes 20 cartons.

40 money tokens are needed for each box, so 320 are needed for a carton. There are 5000 money tokens, so this makes 15 cartons. 12 bonus tokens are needed for each box, so 96 are needed for a carton. There are 1000 bonus tokens, so this makes 10 cartons.

2 dice are needed for each box, so 16 are needed for a carton. There are 300 dice, so this makes 18 cartons.

We only have enough bonus tokens to make up 10 cartons of the complete game, so only 10 cartons can be dispatched. Hence the answer is A.

SECTION ONE — 2012

Question 14: B

D) cannot be correct as there is no orientation in which it could have a hexagon-shape in the bottom-right corner, and a rectangle shape just below the top-left corner, thus it cannot be the same layout as the one in the question.

The other shapes can all be placed in an orientation where there is a hexagon-shape in the bottom right corner and a rectangle shape just below the top-left corner. We now look to see what other shapes these boards have.:

> If A) is placed in the same orientation as the image in the picture, we see that there would be a triangle shape 1 space above and 1 space to the left of the hexagon shape. We can see from the image in the question that this space is uncovered and that there is no triangle shape, so this cannot be the answer.
> In C) we see a triangle shape in the space immediately to the left of the top-right corner. In E) we see a rectangle shape in this space. In the image in the question, we see this space is uncovered and there is neither a triangle nor a rectangle shape in this space, so neither C) nor E) can be the answer.
> If B) is placed in the same orientation as the image in the picture (i.e. rotated 180 degrees), there are triangle shape in the bottom-left corner, and in the space above and to the right of the bottom-left corner. There is also a rectangle shape immediately above the bottom-right corner. In the image in the question, all of these spaces are covered by counters, so this could be the same board as in the question.

Question 15: C

C) correctly identifies a conclusion that readily and logically follows on from the reasoning given in this paragraph.

B) is irrelevant to the argument, whilst A) and E) actually contradict/weaken the argument.

D) is not completely irrelevant to the argument, but it is irrelevant to its conclusion. It offers more explanation of the purpose of punishing directors, but does not directly affect how the argument's reasons lead on to its conclusion.

Question 16: B

The argument's conclusion is that America has high levels of free debate. It's thread of reasons to support this refer to the amount of anti-administration debate in the country following 2004. Thus, if the statement is B) is not true, then it's conclusion is invalid. B) is not stated at any point, so it is therefore an assumption.

C), D) and E) are all completely irrelevant to the argument or its conclusion.

A), meanwhile, actually contradicts the argument, as the argument claims that there was ferocious support of the administration in 2004.

Question 17: C
The argument concludes that diet is responsible for higher cancer rates in Europe. If we accept C) as true, then this suggests something else is responsible, and weakens this conclusion.

B), D) and E) are completely irrelevant to the argument or its conclusions, whilst A) actually strengthens the argument. Suggesting that diet is not a contributing factor towards cancer in areas with low cancer rates reinforces the notion that it may be contributing towards cancer in areas with higher cancer rates (i.e. Europe in this instance).

Question 18: A
From the signal Gordon gets, the bird could be at (21,05), (22,05), (23,05), (20,06), (21,06), (23,06), (24,06), (21,07), (22,07), (23,07) or (22,08).
From the signal Helen gets, the bird could be at (22,05), (23,05), (24,05), (21,06), (22,06), (24,06), (25,06), (22,07), (23,07), (24,07) or (23,08)
From the signal Ingrid gets, the bird could be at (22,07), (24,07)
The coordinates that are possible for all 3 are: (22,07)
Hence the answer is A.

Question 19: D
(100-11) = 89% of men consume more than 6g of salt per day. (100-30) = 70% consume more than 6g of salt per day. If there are equal numbers of men and women then the percentage of the population that consumes more than 6g of salt per day is halfway between this, which is 79.5% or to the nearest percent, 80%. Hence the answer is D.

Question 20: B
The software programmer is organising the program in **descending** size order, and assigning them to the first disc which has enough room for it. In B), the plumber is doing the same thing, organising lengths in **descending** size order, and cutting from the first pipe which has enough length for it.

In E), the plumber is doing the opposite by arranging lengths in **ascending** size order. In D) he is deviating from the method by cutting alternate longest and shortest lengths. In C), he is deviating by simply taking them in the order listed, not organising at all.

In A) he is deviating from the method in the question by cutting the lengths which add up to 12 ft first, a feature which is not seen in the programmer's method.

Question 21: A
The passage concludes that pesticides are a waste of money and it would be better to let other insects control pests. However, if pesticides are more effective than letting the other insects control the pests, then pesticides are not a waste of money, and the argument's conclusion is invalid. Thus, A) is correct.

B), if true, would actually strengthen the argument, but is not essential for it to be valid, so B) is not a flaw. C) is not a flaw because it is not required for there to be a positive benefit associated with using insects to control pests in order for one to suggest it as a solution. Therefore, the fact that the argument does not establish this is not a logical flaw with the argument's reasoning.

D), if true, would also strengthen the argument, so is not a flaw, whilst E) is irrelevant. For a critical thinking assessment, we treat the stated reasons given as true, so it is not a flaw in this context that this fact is not explained or justified.

Question 22: C
The passage describes a situation where there are 2 courses of action, one of which must be followed. One of them cannot be followed, due to dire consequences, so the other must be followed regardless of the drawbacks. C) follows this reasoning. Aspirin or Paracetamol need to be taken, and aspirin cannot be taken due to allergy, so paracetamol must be taken.

A) presents a course of action that would ideally be followed but cannot be. B) presents 2 possible sources of a negative situation, one of which cannot be fixed, so we may as well implement a solution to the other. D) and E) present a 2 courses of action, one of which is simply *preferable*, and not *essential*. None of these alternative answers follow the reasoning in the passage.

Question 23: E
The question describes a principle where policies/initiatives which *help* a situation should not be dismissed because they do not fully solve the problem on their own. Only E) follows this reasoning.

A) contradicts the principle, suggesting a policy be abandoned because it has failed to solve a problem. B), C) and D) are completely irrelevant to the argument, and can be safely ignored.

Question 24: C
Alan bought the jug for £80 and then sold it again for £100 which is £20 profit. He then bought the jug again for £110 and sold it for £120, which is another £10 profit.
Hence the total profit is £10 + £20 = £30. Hence the answer is C.

SECTION ONE 2012

Question 25: D
The crew take off from Rome at 09:05 and arrive in London at 09:55, then take off from London at 10:30 and arrive back in Rome at 13:45. They then depart Rome at 15:05 as that is the next flight and arrive in London 15:55, and then depart London at 16:30 and arrive back in Rome at 19:45.

Hence the total time from first take off to last landing is the time between 09:05 and 19:45, which is 10 hours 40 minutes.
Hence the answer is D.

Question 26: A
We can see that because the two holes in A) are mirror images of each other, and each would be a reflection of the other, if a mirror were placed along the dotted line. Thus, once the paper is folded along the dotted line, they will align onto each other, and the resultant shape will be the holes seen in A), not the hole seen in the picture.

All the other answers, when folded together, will overlap each other's edges, such that a new hole will be seen not resembling either of the holes as seen prior to folding. This will result in a new hole being seen, the same shape as that in the question.

Question 27: B
B) correctly identifies the main conclusion of the passage, which readily follows on from the reasoning given in the passage.
The passage makes no reference to what the result of a referendum would be, whether this would be a positive outcome, or whether the Scottish government receives sufficient funding. Therefore, A), E) and C) are all incorrect.
D), meanwhile, is a reason given in the passage that leads on to support the main conclusion given in B)

Question 28: D
The argument describes how everyone is exposed to asbestos in the environment, and how it is no cause for concern when exposed to it if it is in good condition. D) is the only answer which can be confidently concluded from these reasons.

A), B) and C) are incorrect, as they do not necessarily follow on from the reasons given in the passage. Equally, there is nothing in the passage to suggest that removing asbestos poses more of a threat than leaving it in place, so E) is incorrect.

SECTION ONE 2012

Question 29: C
At no point does the argument state that law-breaking motorists are a significant cause of road accidents, yet it concludes that a substantial reduction in road accidents can be achieved by taking action to prevent law-breaking on the roads. Thus, C) correctly identifies a valid assumption, which is required to be true for the argument's conclusion to be valid.

E) is not an assumption because the passage states that many drivers break the law because the penalties do not act as a sufficient deterrent.
Meanwhile, A), B) and D) do not affect the argument's conclusion, so are not assumptions.

Question 30: B
For the first application, 12 litres is made up which is 1/16 Greatgrass. Hence 0.75 litres or 750ml of Greatgrass is used in the first application. For the second application, 12 litres is made up which is 1/25 Greatgrass. Hence 480ml of Greatgrass is used in the second application.

Hence the total amount of Greatgrass used is 750ml+480ml = 1230ml.
Hence the answer is B.

Question 31: E
There are 200 pupils altogether. 90 got A-C in their mock exams and then again in their actual exams, so this was correctly predicted. Another 60 got other grades in their mock exams and then again in their actual exams, so this was correctly predicted.
Hence for 150 of the 200 pupils, the mocks correctly predicted their GCSE results. This is equal to 75% of the pupils.
Hence the answer is E.

Question 32: A
The shape in A) can be used 4 times to make the trapezium. First, take 2 of the shapes. If we rotate one by 180 degrees, they can then fit together, with the slanted lines joined together, to form a rectangle shape. The other 2 shapes will then fit either side of this, to form the 2 sides of the trapezium.

Triangles can be used to form a trapezium, but 5 are required, so B) and C) are not correct. The shapes in D) and E) obviously do not readily form a trapezium without many being used, beyond the scope of what is expected in the timeframe of a test such as the TSA.

Question 33: A

The passage argues that over time, the intervention group will change their diet, and less strictly follow the healthy diet regime. This will cause their diet to become closer to that of the control group as time progresses. If the control group are also editing their diet according to health information released by the media, this will further narrow the gap in diet between the 2, strengthening the argument that the diets will be similar over time, and will differ less and less.

None of the other possible answers make reference to the notion of the 2 diets becoming more and more similar over time, therefore none of them are relevant to the conclusion of the argument.
Therefore the Answer is A)

Question 34: A

The argument discusses how the police have used excessive force and breached people's liberties, and then goes on to conclude that it would be better if this did not happen and drugs were simply allowed to be sold in nightclubs to keep dealers off the streets. A) correctly points out that the argument has assumed that these are the only options, and neglected to consider other possible ways of stopping drug dealers without such excessive force.

B) is incorrect as the question states that it was a worrying breach of civil liberties. Therefore for the purposes of a critical thinking test we assume that this does constitute a breach of civil liberties.

The other answers do point out flaws in how valid the argument's reasons are, but they do not affect how strongly the reasons support the conclusion. Since we assume that all reasons in a given argument are correct for an assessment such as this, we therefore do not consider these as flaws in the argument.

Question 35: D

D) is the main conclusion of the argument, whilst B) is a reason given in the argument which supports the main conclusion.

A) expresses an opinion which the argument explicitly contradicts, whilst C) and E) are simply facts stated in the passage. None of these 3 answers are valid conclusions from this passage.

SECTION ONE 2012

Question 36: C

Anton's original salary was £29000. Hence his first annual salary increase was £3000 (£2900 rounded up to the nearest £500). Hence his new salary is £32000. Anton's second annual salary increase was £3500 (£3200 rounded up to the nearest £500). Hence his new salary is £35500

- Anton's third annual salary increase was £4000 (£3550 rounded up to the nearest £500). Hence his new salary is £39500.
- Anton's fourth annual salary increase was £4000 (£3950 rounded up to the nearest £500). Hence his new salary is £43500.
- Anton's fifth annual salary increase was £4500 (£4350 rounded up to the nearest £500). Hence his new salary is £48000
- Anton's sixth annual salary increase was £5000 (£4800 rounded up to the nearest £500). Hence his new salary is £53000.
- Anton's seventh annual salary increase was £5500 (£5300 rounded up to the nearest £500). Hence his new salary is £58500.
- Anton's eighth annual salary increase was £6000 (£5850 rounded up to the nearest £500). Hence his current salary is £64500.
- Hence the answer is C.

Question 37: C

In each race, the winner gets 10 points, and anybody not finishing receives 0 points. This means for each race left, a given driver can move 10 points closer to the people above him (i.e. the driver who is winning at that point). So if the number of races X 10 is greater than the difference between a given leader's points tally and the points tally of the leader with the most points, the driver who is trailing can still win the championship.

Thus, we look for a race after which there is a leader who is more points behind the leader in first place than the number of races X 10. After Billard, we see that the number of races X 10 is 40, and that no driver yet has 40 points. So at this point, any leader can win. After Menz, the number of races X 10 is 30. The leader in first place (Damon) has 47 points, whilst the leaders in last place have 24 points. Thus, they can still catch up with Damon.

However, after Fondra, we see the number of races X 10 is now 20, and that David, the leader in last place, is 23 points behind Damon, the leader in First place. Thus, after Fondra, David can no longer win the Championship.

SECTION ONE — 2012

Question 38: C
Since each room must be carpeted with a single piece of carpet, the offcut purchased must be big enough to cut 2 rectangles, one 2.6m by 4.2m, and one 1.8m by 2.2m. There are two ways this can be achieved:
- Firstly, an offcut could be purchased which is at least 2.6m wide, and at least 6m long (long enough for us to cut the rectangle for the smaller room, and still leave enough length to cover the larger room)
- Alternatively, an offcut could be purchased which is at least 4.2m long, and at least 4.2m wide (i.e. long enough to cut a rectangle of 1.8m wide for the smaller room, and still leave enough width to cover the larger room).

As we can see from the choices available, only C) (an offcut of 4.2m by 4.4m) satisfies these minimum criteria. Therefore, the answer is C

Question 39: C
We cannot determine from the passage that removal of these 2 heavy metals would mean the rest of the sludge is safe, so A) is incorrect. Equally, the passage does not say whether it is possible to make sewage sludge safe, or whether it is safer to dump it in the north sea, so D) and E) cannot be reliably concluded from the passage.

B), meanwhile, is in contradiction to the passage, which states that Zinc kills plants at lower concentrations than are needed to harm animals, so this is unlikely to pose a health threat.
The only answer we can reliably conclude from the passage is C)

Question 40: B
D) and E) are completely irrelevant to the argument's conclusions, so are not assumptions. C) is also irrelevant as the argument does not state that financial considerations should be a primary consideration, merely that it is an added bonus. Therefore, C) does not directly affect the argument's conclusions, and is incorrect.
A) is not assumed, as the argument merely states that the chances have receded, not disappeared entirely. Therefore the argument does not rely on assuming that forensic evidence is a foolproof way of determining guilt.

Therefore by process of elimination the answer is B). At no point does the answer explicitly say the statement in B), and if this is not true the argument's conclusion that capital punishment would ensure real justice does not necessarily follow on from its reasoning.

SECTION ONE 2012

Question 41: C
A) and B) are not assumptions as the argument does not claim that it would be suitable to plant trees in urban spaces, it merely claims that it would reduce asthma rates, so other problems with the notion are irrelevant.

E) is irrelevant as the argument does not assign the increase in asthma rates to a lack of trees, it merely claims that more trees would reduce the rates.

D) does not necessarily weaken the argument because we do not have sufficient information on the asthma rates in the question. If these asthma rates being described are relative to the local population, D) would not have any effect on the strength of the argument.

C) would weaken the argument by suggesting that other factors are responsible for the decrease in asthma rates in areas with trees, not the trees themselves. C) implies that the argument has confused cause and correlation, and this weakens the argument.

Question 42: D
If the minute hand travels 3 times as fast down to 6 as it does back up to 12, it must spend a quarter of the hour going down to 6 and then three quarters of the hour going back up to 12.

Hence it reaches the position for half past the hour at quarter past the hour. It then travels the distance which should be half an hour in 45 minutes, which means it takes 1.5 minutes to travel the distance which should be a minute.

Hence it will take 22.5 minutes for it to travel back up to "quarter to the hour" and there will be another 22.5 minutes left in the hour. Hence when the clock shows quarter to the hour, it will actually be 22.5 minutes to the hour.
Hence the answer is D.

Question 43: D
We can work out how long the sun and moon are in the sky together for each town then pick the shortest:
➢ Avonmouth: Between 16:25 and 19:09 = 2 hours 44 minutes
➢ Dun Laoghaire: Between 16:45 and 19:22 = 2 hours 37 minutes
➢ Greenock: Between 16:49 and 19:18 = 2 hours 29 minutes
➢ Hull: Between 16:22 and 18:59 = 2 hours 37 minutes
➢ Leith: Between 16:43 and 19:11 = 2 hours 28 minutes
➢ Liverpool: Between 16:32 and 19:10 = 2 hours 38 minutes
➢ London: Between 16:12 and 18:56 = 2 hours 44 minutes

The shortest of these is Leith at 2 hours 28 minutes. Hence the answer is D.

Question 44: C

This test is most easily answered by looking at each shape in turn, and seeing if it is the same shape as any of the others. If we find that this shape matches any of the others, we can immediately discount both of those shapes from our reasoning. Doing this, we quickly find that A) and D) can be easily identified as the same shape, and that B) and E) can also easily be identified as the same shape. This leaves us with C, which we find does not resemble any of the other shapes. Therefore, the answer is C)

Question 45: E

E) correctly identifies the main conclusion of the argument. This argument follows an unusual structure in that it states the main conclusion right at the start, then gives reasons to support it, rather than the more usual structure of giving the main conclusion at the end.

However, if we look through the other answers we find that B), C) and D) are all intermediate conclusions which if accepted as true, support the statement given in E), which thus serves as the *main* conclusion.

A), meanwhile, is part of the argument's reasoning, and is not a conclusion.

Question 46: C

The reasoning in the passage can be summarised as "A only happens because of B. B is stopping/has stopped happening, therefore A will stop happening". Only C) follows this style of reasoning.
- ➢ E) can be summarised as "A *always* happens if B happens. If B happens, A would happen".
- ➢ B) can be summarised as "A *may* have been caused by B. A started when B started, and stopped when B stopped".
- ➢ D) can be summarised as "If A happens and B did not happen, negative consequence C could happen. Therefore, B must happen".
- ➢ A) can be summarised as "A cannot happen if B happens. B happens, therefore A cannot happen".

None of these other answers follow the same reasoning as the passage.

Question 47: E

B) and C) are irrelevant statements, whilst A) is simply a reason stated in the argument, and is not a principle underlying the argument.

Both D) and E) are valid principles, but only the principle in E) is used in the passage. Since a vast majority of people are right-handed, the argument suggests that buttons be placed on the side which is easier to fasten for right-handed individuals, following the principle in E.

The argument makes no claims about a break with tradition being a reason for the relocation of buttons on women's clothing to the right-hand side.

Question 48: D

The required number of votes to be elected is 127, 50% of the total number of votes cast. Since no candidate has achieved this number, Reeta Marthur will be eliminated and the 2nd preference votes for this candidate re-allocated accordingly. This means there will be 17 votes re-allocated to the other candidates. We can see that no candidate is within 17 of 127, so no candidate can achieve 50% of the vote in the second count.

Therefore we will go to a third count. Even if Wayne Daniels receives all of the votes from Reeta Marthur, he will still be in last place at this point, as the maximum number of votes he can possibly have (if he gets all the re-allocated votes from Marthur) is 35. Therefore, he will be eliminated, and the votes for him re-allocated according to second preferences.

Once these 35 votes are re-allocated, however, the situation can no longer be reliably predicted. We know there will need to be at least one more round of counting, because no candidate is within 35 votes of 127, no candidate can have achieved this number after the reallocation of the 35 votes for Wayne Daniels and Reeta Marthur. However, we cannot predict who will be in last place for the next round of counting.

This is because if Ian McBride receives 10 of the 35 re-allocated votes, he will move above Pedro Gonzales. This would leave Pedro Gonzales in last place for the next round of counting, and he would therefore be eliminated, and his 45 votes re-allocated according to second preferences. This would be sufficient to elevate any remaining candidate into first place. Thus, beyond the elimination of Wayne Daniels, we cannot predict what will happen, and any of the top 4 candidates could win.

SECTION ONE — 2012

Question 49: C

A standard 100W bulb costs £9.60 less than a low power equivalent wattage bulb. Thus, we are calculating how long it will take for a low power equivalent wattage bulb to cost £9.60 less in electricity.

- A 100W will cost 15p of electricity per week if it is on for 20 hours as stated in the question (7.5p per 10 hours multiplied by 2)
- 7.5p of electricity will run a low power equivalent wattage bulb for 50 hours. Thus, in 20 hours a low power bulb will us 3p of electricity per week.

Thus, the weekly saving in terms of electricity cost for 20 hours of run time is 12p. Thus we are looking to calculate how many weeks it will take for a 12p weekly saving to cancel out the £9.60 extra cost of the bulb. 960/12 = 80. Thus, it will take 80 weeks for me to recover the extra purchase cost.

Question 50: E

A) can be easily created by drawing a mirror image along the right-most edge of the shape in the question, then drawing a mirror image along the bottom of the resultant structure.

B) can be created by taking 2 copies of the shape in the image, rotating one by 180 degrees, and fitting them together to form a rectangle shape. Two more shapes can then be added to this, one on the top-right corner and one on the top-left corner, with the inside corner of the L shape on the corner of the rectangular centre, to give the shape in B).

C) is created by forming 2 rectangular shapes, as described for B). One of these is then rotated, as a whole, by 180 degrees, and placed below the other to give the shape in C).

D) is created by drawing a mirror image with a symmetry line going along the left-most edge of the shape in the question. Then a mirror image is drawn with a symmetry line going along the bottom of the shape in the question, to give the shape seen in D).

E) cannot be formed by 4 of the shape in the question.

END OF SECTION

Section 2

Should convicted criminals be allowed to vote?

Introduction
This question is easy to structure around the two elements of this question – criminals and voting. Both need to be explored and clearly defined to answer convincingly.

➢ A criminal can be defined as someone who's action or omission constitutes an offence and is punishable by law.

➢ Voting has two elements; it is both a right – in the sense that individuals within democratic society have a right to say who governs them, and a duty in that it is important that the person voting does so responsibility.

Ideas to explore
➢ One option could be to explore that while the idea of criminality is binary – either you are a criminal or not – the severity of those crimes is not (parking illegally is not as severe as mass murder), and therefore a convincing argument could be made to allow certain criminals – i.e. those who commit less bad crimes – to vote.

➢ A good approach would be to point out that while voting is a right, it is an alienable right in that it can be denied to some people within society. One option could be to look at other groups in society who are not allowed to vote – for example those under the age of 18 – and explore the rationale behind the decision. In this case those under 18 are deemed to not be sufficiently mature politically

Arguments against
➢ One potential argument is that criminals gave up their right to vote. If a person optionally chooses to commit a crime, they are branding themselves criminals. Criminals should not be allowed to vote. Why should the leaders of a democracy be partly chosen by those who would seek to ruin and harm those around them?
➢ A democracy does not function as a democracy when the troublemakers are making the decisions.

Arguments for
- One of the mains functions of prison is to rehabilitate prisoners who have done crimes in the past and to allow them to be functioning members of society. Not allowing them to vote would alienate these criminals from society and thus not allow them to become functioning members of society.
- A convicted criminal or not, criminals are still members of our society. To say they no longer deserve a voice is not only inconsiderate but selfish. Just because their opinion may not align with yours, does not mean that it should be stripped from them. Prisoners are put under the protection of the government and therefore have the full right to be able to have a say on who protects them.

Conclusion
- Restate your position and summarise the main arguments you have used to defend it.
- Link your conclusion back to the question by referring to the definition of 'criminal, what does it mean for someone to be a 'criminal' in regards to their place in society?
- The key thing in this question is the idea that voting is an important duty that cannot be given to just anybody - can society can trust someone who has violated the laws of society with having a say in who governs it?

Does a country's ideal political system depend on its level of economic development?

Introduction
- This question requires an answer that explores and defines different political systems and different levels of economic development. Unlike other questions, both the definition of political system and economic development are complex and would be very difficult to define in full within the time allocated. As a result, some simplifications must be made.

Arguments
- What is meant by ideal political system? 'ideal' is a very subjective term – ideal for who? Try to think of some examples of an ideal political system for different situations. A political system which results in an equal and fair income distribution would be ideal for the majority, whereas a system which allows for a militarised industrial complex and a large number of loyal soldiers would be ideal for a country engaged in a war.

> Is there anything unique to the level of a country's development which determines what is the ideal political system for it? It is possible to argue for that a country which is very undeveloped economically, a political system which aids development and an increase in wealth would be ideal in order to reach minimum levels in health, education and income. In this sense, it could be argued that the ideal political system could depend on economic development. However, once these minimums are reached, there is little to link an ideal political system to development.

Real life examples

Economic growth may be affected by a great deal of extraneous factors independent of a political system (international economic environment; industrial cycles; geopolitical factors, both permanent and temporal - including natural resources, navigable rivers, etc...). Use real life examples of economic development. For example:

> France was able to relatively well economically since WW2, but how much of that can be attributable to USA taking on the costs of allowing France to be safe from both an invasion by Germany AND by USSR, no matter what French thought of NATO?

> China's political system change had less to do with the economic growth than both the absolute zero level that the economy started with when Deng Xiaoping started; AND the fact that the western countries were at an economic point of wishing to outsource industrial production. If China was on its own, it would not have seen the same growth even remotely.

> Other examples: Democracies such as the United States is doing better financially than some Authoritarian countries such as North Korea. However some Authoritarian countries such as Libya (now a Democracy?) and Saudi Arabia have profited well due to natural resources. Since most Non-Democratic Republic countries are in the Middle East/ Africa, they have profited well from oil – what effect does this have on the question?

> If you consider that Authoritarian leaders can spend all the money they want to, it might make an Authoritarian government less economically secure. Also considering that leadership can't be changed even if they are destroying the country. Mainly though, economic growth depends on location and allies.

➢ Also, if an Authoritarian government is unstable and are at war with another country, or rebels due to their leadership, they might be spending the majority or their money on their army. In a democracy the budget is regulated by a congress, or parliament so that the right choices can be made

Conclusion
➢ This question is difficult due to the complex nature of the subjects covered. As a result, it is vital that candidates take extra effort to make their definitions, arguments and examples very clear throughout.

➢ Restate your position and summarise the arguments you have used to support it.

➢ Link your conclusion back to the question by referring to your definitions of a 'political system' and also 'level of development'. What, if anything links the two?

Should governments only fund scientific research if it is of direct benefit to society?

Introduction
Explore the ideas of government funding and scientific research:
➢ What is the idea behind government funding? It is plausible to define 'correct' government funding as that which directly benefits society.
➢ What is the nature of scientific research? What does it try to do?

Arguments for
Government funding and public funds
➢ Government funding comes from taxation, which itself comes from members of society. It is, therefore society's money and should be used for their benefit.
➢ Play with the question – should government spending as defined above be used to fund something which results in the benefit of an elite number of individuals? Clearly not – public money should be used for public benefit.

Scientific Research

- A key issue here is that scientific research is not predictable in terms of results due to its experimental nature.
- Some of the most beneficial scientific breakthrough occurred by accident – for example the discovery of penicillin happened purely because a scientist who was trying to discover something else left a Petri dish out by mistake. Equally, some research projects which had good intentions have resulted in discoveries that are detrimental to humanity – such as for example Alfred Nobel's discovery of TNT. These ideas can be extrapolated into arguments, as below.

Arguments against

- Research should not have to be economically justified, as academic research is often an issue of "what happens when…". The results are often not predictable, but they often have a huge beneficial impact.
- Computers, television, modern vehicles and modern building advances, have all been impacted by university (public) research. A world where research couldn't exist without a business justification would be without these amenities.

- More examples: the current hard drive in your computer probably benefited from research done in computer labs for NASA. NASA wanted high resolution cameras on a couple satellites, but had to have a hard drive that wrote quickly enough that the steady stream of data received wouldn't be lost. Was there a commercial value at the time? No, just a scientific one. But computer manufacturers and users now pay royalties for the research done 20 years ago. Indirect benefits to the scientific community can lead to direct benefits to society in the future. Using examples like this will help make your argument more convincing.

Conclusion

- While in theory the government should fund only that which is in the benefit of society, in the field of scientific research, it is impossible to know exactly which research projects will have positive (or negative) results. Therefore, in reality, the government should, as a general rule, fund projects which can reasonably expected to produce positive results for society, but, as per the reasons above, should not rule out funding those with less of a direct benefit.
- You may wish to come to a decision either way, or it is equally fine to sit somewhere in the middle, so long as this is fairly justified.
- One option could be to conclude with a normative policy guide in response to support whatever side you side on.

Could a robot ever think like a human?

Introduction
This argument centres on what you understand of the concept of humanity. Make sure that you clearly outline a definition and stick to it throughout the essay.

Real life application and examples make answers much more impressive. In this case, the Turing test is an example of a useful theory to base your argument around:

- The Turing test is a test of a machine's ability to exhibit intelligent behaviour equivalent to, or indistinguishable from, that of a human. Alan Turing proposed that a human evaluator would judge natural language conversations between a human and a machine that is designed to generate human-like responses. Think about the implications this test has for the question.

Argument for
- While the brain is incredibly complex, and is able to perform X million calculations per second, looking at the development of computer technology over the past 50 years, it is just a matter of time before a computer with a comparable computing ability is designed.
- In theory, any neural process can be reproduced digitally in a computer, even though the brain is mostly analogue.

Argument against
- What it is to be human goes beyond just the ability of our brains to make millions of calculations per second – there is an emotional and sensitive side to humanity that computers will never be able to replicate, no matter how complex computers become.
- While sensory feelings like heat, cold or pain could easily be felt from the environment if the machine is equipped with the appropriate sensors, this is not the case for other physiological feelings like thirst, hunger, and sleepiness. These feelings alert us of the state of our body and are normally triggered by hormones. Since machines do not have a digestive system or hormones, it would be downright nonsensical to try to emulate such feelings.

- Emotions are too complex for robots: emotions do not arise for any reason. They are either a reaction to an external stimulus, or a spontaneous expression of an internal thought process. For example, we can be happy or joyful because we received a present, got a promotion or won the lottery. These are external causes that trigger the emotions inside our brain. The same emotion can be achieved as the result of an internal thought process. If I manage to find a solution to a complicated mathematical problem, that could make me happy too, even if nobody asked me to solve it and it does not have any concrete application in my life. It is a purely intellectual problem with no external cause, but solving it confers satisfaction.

Counter Argument
- Is there anything inherently un-computable about the emotional side of humanity? Emotions are, after all, just another of one of the many computations that the brain makes – it is simply the receiving of data and acting accordingly.

Conclusion
- This question boils down to whether you agree with the idea that what it is to be human is merely having a complex piece of matter in your head – your brain - which is able to perform a large number of calculations.
- Restate your position and summarise the main arguments you have used to defend it.
- Link your conclusion back to the question by referring to your definitions of 'robotics' versus 'humanity', and what about them, specifically, makes them different or similar.

END OF PAPER

Afterword

Remember that the route to a high score is your approach and practice. Don't fall into the trap saying that *"you can't prepare for the TSA"*– this couldn't be further from the truth. With knowledge of the test, time-saving techniques and plenty of practice you can dramatically boost your score.

Work hard, never give up and do yourself justice.

Good luck!

About Us

Infinity Books is the publishing division of *Infinity Education Ltd*. We currently publish over 85 titles across a range of subject areas – covering specialised admissions tests, examination techniques, personal statement guides, plus everything else you need to improve your chances of getting on to competitive courses such as medicine and law, as well as into universities such as Oxford and Cambridge.

Outside of publishing we also operate a highly successful tuition division, called UniAdmissions. This company was founded in 2013 by Dr Rohan Agarwal and Dr David Salt, both Cambridge Medical graduates with several years of tutoring experience. Since then, every year, hundreds of applicants and schools work with us on our programmes. Through the programmes we offer, we deliver expert tuition, exclusive course places, online courses, best-selling textbooks and much more.

With a team of over 1,000 Oxbridge tutors and a proven track record, UniAdmissions have quickly become the UK's number one admissions company.

Visit and engage with us at:
Website (Infinity Books): www.infinitybooks.co.uk
Website (UniAdmissions): www.uniadmissions.co.uk
Facebook: www.facebook.com/uniadmissionsuk
Twitter: @infinitybooks7

www.ingramcontent.com/pod-product-compliance
Lightning Source LLC
Chambersburg PA
CBHW071402290426
44108CB00014B/1657